Energy or Extinction?

Energy or Extinction?

The Case for Nuclear Energy

FRED HOYLE

HEINEMANN
London

Heinemann Educational Books Ltd
LONDON EDINBURGH MELBOURNE AUCKLAND TORONTO
HONG KONG SINGAPORE KUALA LUMPUR NEW DELHI
IBADAN NAIROBI LUSAKA JOHANNESBURG
KINGSTON/JAMAICA

ISBN 0 435 54430 6

First published 1977
Reprinted 1977 (twice), 1978

Published by
Heinemann Educational Books Inc.
22 South Broadway, Salem
New Hampshire 03079

and

Heinemann Educational Books Limited
48 Charles Street, London W1X 8AH
Printed Offset Litho and bound in Great Britain by
Cox & Wyman Ltd, London, Fakenham and Reading

Foreword

by

Sir Alan Cottrell, F.R.S.

formerly Chief Scientific Adviser to H.M. Government

This is an important book which I hope will be read and thought about deeply by politicians and by everyone concerned with the future of western democratic society. It is about **energy**: about the alarming prospect that oil will soon run out and not be replaced by anything else. It shows that – contrary to an influential belief – we do *not* have time, that there is *no* practical alternative to nuclear energy, and that western decision makers have been frightened into immobility in their nuclear energy policies by a well-orchestrated campaign which has marched under an 'environmentalist' banner but yet has a clearly identifiable political basis.

All Fred Hoyle's writings are brilliant and strikingly original. This is true vintage, but he also brings to it a passionate intensity of feeling about the vulnerability of western society and of the threat to our children's future. Written for the non-scientist, it explains the scientific, economic, and political backgrounds to the world's energy and materials resources; and it brings refreshing commonsense to bear against the claims for new non-nuclear energy sources and the hysteria of the anti-nuclear environmentalists.

The book makes two major points, hardly to be found in any other book on the energy crisis. First, that some of the anti-nuclear campaigns are politically inspired as a means of weakening the West. Second, that nuclear breeding – which is now the only sure way to an energy-unlimited future – does not necessarily require fast breeder reactors; it can also be accomplished in reactors such as the Canadian CANDU type, which are technically straightforward and well-proven. It is surely a scandal that the U.S. and the U.K. have ignored this attractive alternative for so long.

1977 Alan Cottrell

Contents

I wish to thank my wife, Barbara, and my son, Geoffrey, for
their considerable assistance in the preparation of this book.

1 The Anti-nuclear Environmentalists

In a number of writings I have myself welcomed the concern now felt by many people for saving birds, trees, natural beauty, the whale, and the Atlantic salmon. It is unfortunately the case however that there are always individuals waiting around to seize on any worthwhile popular movement, political animals who manipulate such movements for their own ends. Half a generation ago we described ourselves as bird-watchers, or just plain out-of-doors types – mountaineers and the like. But now we have all suddenly become 'concerned persons', just the kind of word-label which political manipulators use to make us forget precisely what explicit issues are troubling us. For only by persuading us to forget explicit issues can a popular movement be perverted to serve the ends which the political animals really have in mind. In this short first chapter I shall explain what I believe the real motive of such environmentalists to be.

I believe the motive to be connected with the Soviet Union, and with a world struggle for energy. The economic system of the Soviet Union has many disadvantages. It is not technically very innovative. It does not produce consumer goods efficiently. But the Soviet system is not entirely besotted, as the Western democracies are, with the illusory importance of paper money. The crucial principle, that energy is more important than money, would be more easily seen in the Soviet Union, perhaps very easily seen, than it is here in the West. So I would expect the geographical distribution of world energy reserves (of the kind that dominate our present day energy use) to be of great interest to a Russian.

The geographical distribution of coal is shown in Figure 1.1. The dominance of Soviet reserves is manifest.

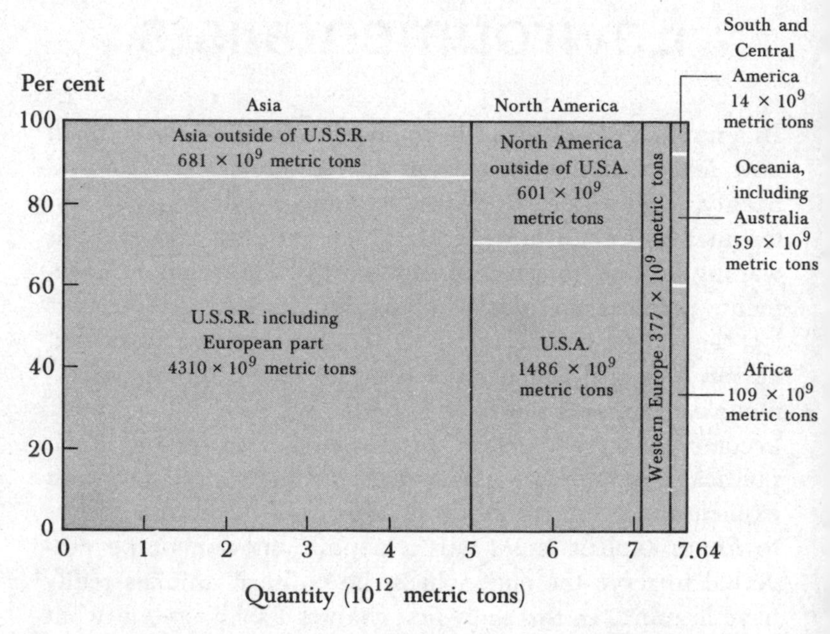

Figure 1.1 The geographical distribution of coal

The distribution of the world's major oil fields is shown in detail in Figure 1.2. If you were Russian, you would surely take careful note of that great crescent containing nearly 70 per cent of world oil reserves which starts in the U.S.S.R. and sweeps through the Middle East into North Africa. You would see the strategic importance of Israel nestling there between the horns of the crescent, Israel the one firm base from which your western enemies could operate to prevent you from exercising direct physical control over the Middle East. You would also notice the great bulk of Africa around which tankers from Europe and North America must go to reach the oil fields of the Middle East, and you would realize that control of the western coastline of Africa would permit you to cut those tenuous shipping

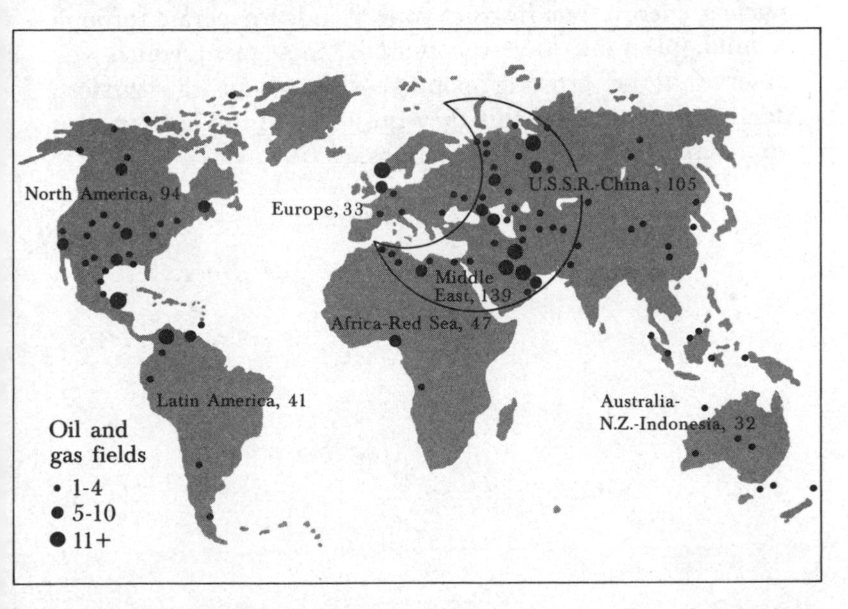

Figure 1.2 The distribution of major oil fields (from Arthur A. Meyerhoff, *American Scientist,* September 1976, reprinted by permission)

lanes. So you would set your many vociferous friends throughout the world howling and baying for the blood of Israel and South Africa. And to develop your muscle you would expand your navy, especially its submarine complement. You would also set yourself to exploit the many political troubles to which the continent of Africa is endemic. Believing in the all-importance of energy, you would scent victory in the world struggle. Marxists never stop talking about the world struggle – they believe in that too.

The fly in this otherwise smooth ointment, which in your Russian guise you have prepared, is nuclear energy. Your western enemies have a powerful nuclear technology, to a degree where it would not be outstandingly difficult for them to obtain access to all the energy they need. Evidently then, you start your vociferous friends in the West baying against nuclear energy. You instruct your friends to operate through a mild, pleasant, 'save-the-animals' movement which you observe to be growing popular throughout the western democracies. And all this they do, right to the last letter of your Kremlin-inspired instructions.

2
Stars and Atoms

The world around us is obviously very complicated, and just as obviously full of marvellous things. A small seed planted in the ground grows into a tall tree. Distant stars cover the sky at night like a vast display of twinkling gems. Storms sweep over the ocean, and volcanoes rumble and roar as they burst out in cascades of molten lava. A piece of dull stone, if we treat it in the right way, can be changed to a globule of shining metal, and day-by-day life itself is made possible by the warmth of the Sun. So complicated is the world that it is remarkable that man has succeeded in making so much sense out of it. The outcome of our studies is indeed often more astonishing than our first guesses. The Greeks believed the Sun to be a fire, like the fire in a brazier, that was somehow hauled across the sky each day by a god-like agency. Today we know the Sun crosses the sky once each day because the Earth on which we live turns around like a spinning ball, sometimes causing us to face the Sun and sometimes to turn away from it. And today we know that the Sun is a vast nuclear reactor, and that the energy we receive from it comes from nuclear energy.

In this chapter I am going to discuss a little of the history and the nature of materials which make up the soil, the ocean, and the air. The information to be obtained from this seemingly abstract study turns out to be remarkably relevant to many of our social and economic problems.

If you were to take any hunk of material and were to hammer it hard enough you could break it into pieces. And if you were to hammer the pieces you could break them into still smaller pieces. Imagine you would always equip yourself with the right kind of hammer. Could you go on with this dividing into pieces for ever? The Greeks pluralists guessed not. A stage would come, they guessed, when no further division was possible, no matter how powerful the hammer. The bits into which matter had been divided at this stage would be the smallest possible bits, which the Greeks called **atoms.**

Today we are still asking the same question, and today many scientists still give the same answer as the Greek pluralists. But no one yet has reached the supposedly indivisible pieces! Scientists in the nineteenth century thought they had though, and, following the Greeks, they gave the name 'atom' to any piece which they believed to be indivisible. The nineteenth century use of the word 'atom' became so widely diffused in scientific literature that modern scientists have felt it best to keep to this usage, and then to invent new names for the bits into which the atoms of the nineteenth century can nowadays be divided. Over the first half of the present century the bits into which atoms could be divided were called electrons, protons, and neutrons. In the past few years still more powerful hammers have been constructed (hammers which physicists call 'accelerators'), and with them it has been shown that protons and neutrons are themselves made up of still more elementary pieces, pieces with the name **quark** (a word chosen to be different from any word used previously in science).

But let us return to the atoms of the nineteenth century. They were found to come in about 90 varieties. Many have familiar names; carbon is one, nitrogen another. The oxygen we breathe is another. The iron in our blood, and the calcium in our bones, are others still.

There are of course many more than 90 different kinds of substances to be found in the world around us. This is because the 90 kinds of atom can join together in clusters called **molecules**; each kind of molecule characterizes a 'substance'. The simplest molecules are those with clusters of just two atoms, like the molecule of common salt which has one atom of sodium and one atom of chlorine. Complex molecules have clusters with very many atoms; the green chorophyll in the leaves of plants is a substance whose molecules each contain more than 100 atoms, while molecules containing as many as 10 000 atoms make up the genetic material in the cells of plants and animals.

The hammers with which scientists in the nineteenth century were able to equip themselves were powerful enough to break up molecules into their constituent atoms, but not powerful enough to break up the atoms themselves. Mostly the hammering was done simply by supplying heat to a material (hence the ubiquitous bunsen burner) which caused the molecules to collide more violently with each other. With enough heat even apparently refractory materials like limestone can be broken down into bits. Limestone is broken into carbon dioxide and lime in a kiln. Both carbon dioxide (CO_2) and lime (CaO) are still molecules, but these molecules can themselves be broken into their constituent atoms, carbon (C), oxygen (O), and calcium (Ca), by a still greater heat.

Such 'chemical changes', in which atoms are shuffled around from one kind of molecule to another but without the atoms themselves being changed at all (as compared to 'nuclear changes' when the atoms are changed), were by no means confined to the laboratory. Chemical industries, concerned both with breaking up old molecules and with forming new ones, began to grow up in the nineteenth century because the substances formed out of the changed molecules were found to be useful – for example lime could be used by the farmer to sweeten sour land. Even from our modern point of view the events which go on around us arise overwhelmingly from chemical changes. As plants grow, complex molecules are being built up from simpler molecules which the plant takes in from the ground and from the air. Sunlight falling on the plant allows these complex molecules to be formed; the energy of light is transformed into chemical energy. And with our cars the engine causes the car to move because of an energy transformation, from chemical energy first to heat and then to motion, which has its origin in the combination of gasoline molecules with oxygen from the air.

Now the Earth has not existed for ever. It was formed along with the other planets about 4600 million years ago, probably in the process which led to the formation of the Sun itself as a star, by a condensation of gas within a cloud like the well-known nebula in the constellation of Orion; the 90 or so different kinds of atom therefore came a long time ago from a gas cloud, in which they had the relative abundances shown in Figure 2.1. Quite a bit needs to be said to understand the meaning and the importance of this Figure.

It is useful to classify the different atoms by their masses. For this we take the lightest atom, hydrogen, to count 1. Then a carbon atom counts about 12, because in round numbers carbon is 12 times heavier than hydrogen. And an iron atom counts about 56, because also in round numbers iron is 56 times heavier than hydrogen. Using these 'atomic masses' we can order the different kinds of atom as on the bottom scale of Figure 2.1; we can also order the relative abundances of the different kinds of atom in our solar system on the left-hand scale, so that each kind of atom then yields a point in the Figure. Notice that this left-hand scale is *logarithmic*. Each unit step of the scale corresponds to a whole order of magnitude, a factor 10, of abundance. Atoms of hydrogen are therefore about 1 million million (10^{12}) times more abundant in our solar system than the varieties of atom with the least abundances. This difference is so

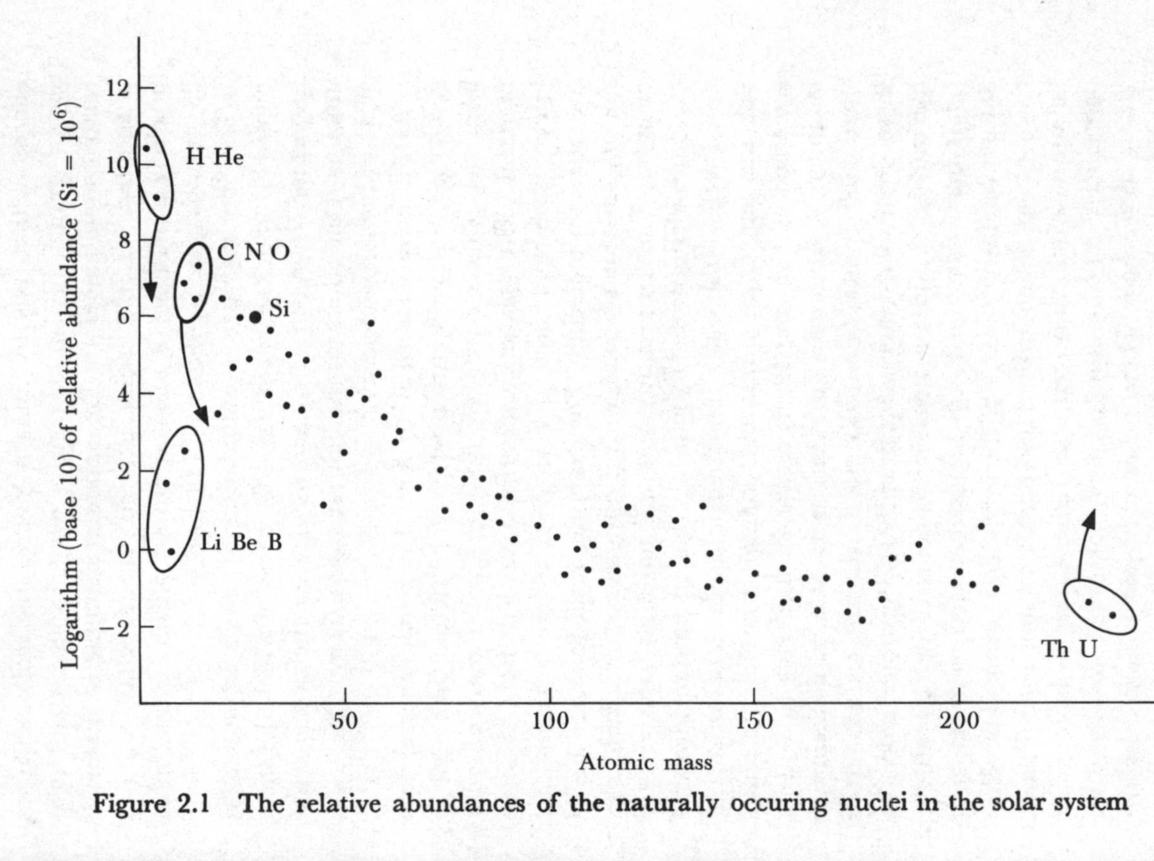

Figure 2.1 The relative abundances of the naturally occuring nuclei in the solar system

enormous that only by using a logarithmic scale can it be shown graphically in a concise way. Notice that these relative abundances are for a standard sample of material, a standard sample being defined as one that contains precisely a million (10^6) silicon atoms; silicon (atomic mass 28) thus has exactly the level 6 in Figure 2.1.

The Sun has relative abundances of atoms as plotted in Figure 2.1; but the Earth has differences of which the most important are indicated by arrows, one for the two heaviest atoms and the others for the lightest atoms. These differences arose in the following way.

As the gases that went to form the whole solar system condensed, a stage occurred where the fraction that went to form the planets separated from the main mass of solar material. This planet-forming gas cooled from a high temperature, and as it did so most of the atoms joined together into molecules. For example, most (but not all) of the oxygen atoms joined with hydrogen to form molecules of water. The temperature was still too high, however, for the molecules of water to join together to produce drops of liquid water – the water remained in the form of hot steam. But the temperature was not so high that all molecules remained in a gaseous form; molecules like silica (one atom of silicon, two atoms of oxygen) were able to condense as small solid bodies. Gradually the Earth accumulated from these small solid bodies. Meanwhile still gaseous molecules like carbon dioxide and water evaporated away, with the result that when the Earth at last emerged as a planet it did not retain very much of the gases. So it came about that those atoms which existed mainly as vapour – hydrogen, helium, carbon, nitrogen, all light atoms – have great deficiencies on the Earth when compared to the solar system as a whole. These deficiencies are indicated by the arrows at the left of Figure 2.1.

Events on the Earth itself occurring over the long history

of our planet have also changed the abundances of atoms in the surface rocks. Atoms which form comparatively volatile substances, sulphur for example, have become deficient in the surface rocks, while some atoms, barium for example, have developed considerable excesses. Of later importance to man, most of the uranium and thorium atoms – atoms

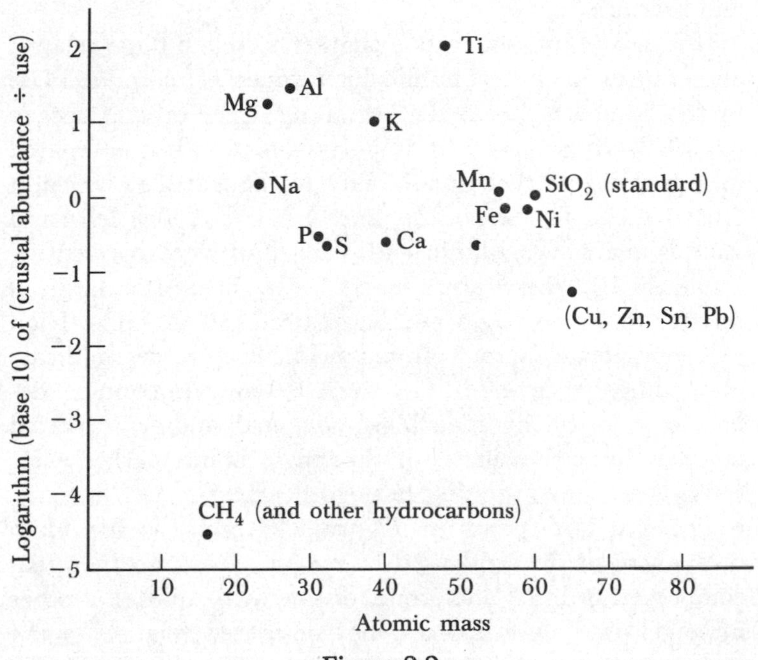

Figure 2.2

from which *we can derive energy by nuclear methods* – went, for reasons which are not entirely clear, into the top few hundred kilometres of rock, near to the surface of the Earth. This enhancement is shown by the arrow at the right of Figure 2.1.

Oil, coal, and gas – the hydrocarbons – all have molecules made up from carbon and hydrogen (in ways varying from one hydrocarbon to another). The fact that the Earth has great relative deficiences of both carbon and hydrogen is the reason why there is so little of the hydrocarbons here on the Earth. This deficiency of hydrocarbon-forming atoms, taken with the relative enhancement of atoms of uranium and thorium should be a clear invitation to us, a literal sign from the heavens, to replace our present energy dependence on the hydrocarbons by nuclear energy derived from uranium and thorium.

It is also of interest in this chapter to relate these relative abundances to the economic importance of materials. Use by the United States at the beginning of the present decade (ca 1970) of economically important materials is compared in Figure 2.2 to their availability in the Earth's outermost crustal rocks, in the oceans, and in the air. The left-hand scale is again logarithmic with each unit step representing a factor 10; these abundances have been standardized relative to the widely occurring material silica, SiO_2. High values indicate ample abundances, low values indicate impending deficiencies. The overwhelming impending deficiency is for the hydrocarbons – oil, coal, and gas – plotted together for convenience in the single point marked CH_4 (CH_4 is methane, the simplest hydrocarbon).

The standard point in Figure 2.2, SiO_2, is the main component of the sand used in cement. It is also the main component of brick and stone. So the availabilities of other materials in Figure 2.2 can be considered relative to the basic component of building materials. Relative to their use (i.e. the present economic demand), sodium (Na), manganese (Mn), iron (Fe), and nickel (Ni) are about as common as basic building materials. Relative to their use, magnesium (Mg), aluminium (Al), potassium (K), and titanium (Ti) are more common; and relative to their use, phosphorus (P), sulphur (S), calcium (Ca), chromium (Cr), copper (Cu), zinc (Zn), tin (Sn), lead (Pb), and the hydrocarbons are all less common, the hydrocarbons enormously so.

A whole world of economics and politics is contained in these statements about the natural state of things. 'Oil politics' is the most immediate example, ranging from embargo threats to 'obscene profits'. Developing nations are currently seeking through UNCTAD, a United Nations organization, to obtain what they call a fair price for their commodities, among which are the metals Cu, Zn, Sn, and Pb. Nobody is seeking a 'fair' price for sodium, because sodium is common enough to be available to everybody. A 'fair' price really means a scarcity price. The developing nations feel they should be receiving a scarcity price, just as the oil producers are receiving a scarcity price for their hydrocarbons.

The reason why Pb, Zn, and Cu particularly are not selling on world markets at prices which their real physical scarcity would almost seem to guarantee is that excess production, of Cu especially, is being dumped on world markets because the developing countries who produce Cu cannot trust each other to maintain a sensible marketing policy. So what the developing countries are demanding is that a cartel be organized for them.

Bauxite producers should take warning from Figure 2.2. Bauxite is a mineral with a high aluminium content. But vast quantities of Al exist in ordinary clay. Although more energy is required to extract aluminium from clay than from bauxite, should the demand for a 'fair' bauxite price go too far, aluminium companies will simply turn to clay.

While P, S, Ca, and Cr have very similar abundances relative to their use, these **elements** (as atoms are often called) differ markedly in relation to their economics and politics by reason of their geographical distribution and their individual chemical and physical properties.

Calcium is widely distributed in the form of limestone and is also found on ocean beaches in the useful form of shell-sands. Consequently no nation has a grip on the world's calcium supply. Nobody has ever gone to war for calcium, at any rate in modern times. The underabundance of sulphur relative to use is also a low-key situation, because the high volatility of sulphur and its compounds permits sulphur to be easily recovered as a by-product of many industrial processes. The few per cent of sulphur found in many iron-bearing ores can be recovered in this way. Nobody has gone to war for sulphur either.

Nobody I think has actually gone to war for phosphorus, but Morocco was recently prepared to do so. It will be recalled how a band of Moroccans was assembled at the northern frontier of the Spanish Sahara. Because Spain preferred not to resist the march south of this band there was no war and Morocco was permitted to seize the phosphate resources of the Spanish Sahara. Since no important political relationships were involved the industrialized countries were not interested in the fate of the indigenous population, and because there was nothing which the developing countries could grab for themselves they too were not interested. The United Nations therefore thought it unnecessary to refer to its own Charter on the rights of self-determination of indigenous populations.

The case of chromium is different indeed. Chromium is critically important in the making of speciality steels and is therefore a highly strategic material. As if in recognition of this fact, spiteful fate, which Homer knew all about, has arranged that essentially all of the world's chromium-bearing ores (chrome) are located either in the Soviet Union or in Rhodesia and South Africa. This places American politicians in a tough dilemma.* If they come out in support

* The dilemma has been solved temporarily in an ingenious fashion. By importing excess supplies from Rhodesia and South Africa over a number of years, the United States has built itself a stockpile of chrome. This now permits American politicians to strike a moralistic posture on the problem, which they will be able to maintain so long as the stockpile lasts.

of South Africa, the black vote will surely turn against them at the next election; if they oppose South Africa, the collapse of the white regime there could place the strategic supplies of chrome in serious jeopardy. And by a like token the Soviet Union and the other Warsaw Pact countries, together with their many friends among the developing countries, step up their pressure on South Africa. While control of the world's chrome supply would not in itself give control of the world to the Soviet Union (as dominance of the world's energy supply surely would), it would certainly be an excellent move(!!) on the world's political chessboard.

3
Energy

Energy is more important than money. We tend to think otherwise because we use money to buy energy. But if no energy were available to be bought the situation would be changed drastically for it is really energy, not money, which gets things done, and this is what life is about. It needs energy for you to move around, for your blood to flow, for you to see, to feel, to think. So the statement that a society without energy is not merely inconceivable but truly impossible is correct absolutely. Yet the odd thing about energy is that you never use it up. In fact you don't consume it at all, you only transform it from one kind of energy to another. These transformations are studied in the branch of science known as thermodynamics.

Energy exists in many forms – energy of motion, energy of position (a skier gains motion by falling downhill), electrical energy, energy of light, chemical energy, nuclear energy, energy of heat. The essential concept is that these various forms are not equally transformable, one form into another. The heat energy present in the average house would be sufficient in quantity to propel all its inhabitants to the Moon, if it could be used in that way, which fortunately for the household (and unfortunately for NASA) it cannot.

Heat is the least useful form of energy. In very cold weather when you have a desperate need for heat you might think otherwise. With other energy forms there is always the option of turning them into heat, but when you start with heat the reverse option is not *fully* available to you. To understand something of the nature of heat, start with a football kicked into motion by a player. The kick transforms chemical energy in the player's leg muscles into the energy of motion of the ball. This energy of motion is not heat, however, because all the atoms of which the ball is made up are going the same way. But if in some manner the atoms could have their directions of motion changed, some moving forward, some backward, some sideways to the left, some sideways to the right, some up, some down, the usefulness of their energy of motion would have been reduced (indeed the player would be considerably mystified, and the football game would come to an instant stop). The energy of motion would then be on its way to going into heat. The reason a collection of atoms moving in random directions is less useful than the same atoms all moving together is that this process of *randomization* cannot be reversed – the footballer's kick can be converted into heat, but the heat could not be transformed back again into the footballer's kick.

But to say heat energy is not as useful as the other forms of energy is not to say that heat energy is entirely useless. If the atoms of the football after having their motions randomized were to pass some of their energy to the environment (the air or the ground), a fraction of the original energy could be put back again into the motion of a second football. But no matter how ingenious the engineer who contrived the practical details of such a process, the speed of the second football would not be as fast as the first one. Energy of motion would irretrievably have been lost, with the balance of the energy going irrecoverably into the environment. So it always is when heat energy goes into the other energy forms. Nature exercises a tax on the transaction.

The most appropriate date for defining the beginning of modern industry would seem to be 1712, the year Thomas Newcomen succeeded in operating a steam engine for converting heat energy into the energy of motion of a pump.

Nature exercised an energy tax of more than 99 per cent on poor Thomas (but still not as bad as the 102 per cent money tax which the Swedish government recently exercised on one of its citizens). Engineers then started looking for energy tax loopholes. By 1774 John Smeaton had slightly whittled Nature's tax down to about 98 per cent, by 1792 James Watt had got it down to 95.5 per cent, and by 1830 or thereabouts Arthur Woolf and Richard Trevithick in England and Oliver Evans in the United States had pushed it down to about 85 per cent. Modern steam engines operate at a tax of about 70 per cent. There is hope of an eventual reduction to about 60 per cent but nobody expects that it will be possible ever to do much better than that.

In contrast to Nature's high tax on the use of heat energy, other energy forms are fairly readily transformable one into another. It is true that a tax is always levied on every transformation, with the tax taking the form of useless heat transferred to the environment, but the tax rates in such cases are very much lower. Energy of motion is convertible into electrical energy with the aid of a device called a dynamo, and the reverse happens with an electric motor. The taxes in these cases come from heat produced in wires and in bearings. By attention to the manufacture of the bearings engineers have been able to pull the mechanical 'losses' down to a fraction of a per cent in some cases – the bearings of a modern telescope weighing several hundred tons are so exquisitely made that you may move the whole telescope with a finger, so much has friction been reduced. Friction, the rubbing of surfaces, is a very common cause of the production of unwanted heat.

Chemical energy is remarkable stuff, particularly in the way that the chemical energy of the food we eat is transformed inside our own bodies. The first step is to transform the chemical energy of the food into more versatile forms of chemical energy. Then these more versatile forms are used for repairing damage to the body, for conversion into energy of motion through the action of our muscles (thus permitting us to move around and search for more food), and for conversion into electrical energy thus permitting our brains to function (so that we can think how best to search for the food).

In each of these bodily transformations heat is produced, particularly by the muscles and by the brain. Eventually the heat winds up in the environment, where it finally becomes useless to us. But on its way to the environment the heat is very useful to us – it keeps us warm. And it is used in a precise way – to keep the brain at constant temperature, which is a condition essential to being able to think clearly. (Occasionally the multiplication of a virus or bacterium in your body interferes with the chemical transformations, causing more heat than usual to be produced. This excess heat gives you a 'temperature', literally causing the temperature of the brain to rise. A rise of only a few degrees leads to obvious malfunctions. It makes you delirious.)

The amount of heat we produce could hardly have been better judged. Without clothes it is possible to live with reasonable comfort in an equatorial belt from latitude 30° S to 30° N, which comprises half of the area of the Earth's whole surface. If the heat production had been more, sufficient for us to live unclothed in the smaller areas around latitude 50°, say, then people living in the larger equatorial belt would have overheated. Notice, however, that in balancing our heat production to permit us to live without clothes in the largest possible area, Nature overlooked the things that man's quickly developing brain would later cause him to do; first to want to move into the temperate zones to hunt large animals feeding on the grasslands there; and at a still later stage to use the temperate zones for the growing of crops. But this apparent oversight didn't matter logically because a brain which develops far enough to want to hunt large animals must necessarily have developed far enough to invent clothes. It is interesting that Nature (in its biological developments) also overlooked what Nature was doing to the detailed geography of the Earth. On the Earth most of the technically advanced material resources happen to lie outside the tropics. The right-hand (biology) did not know what the left hand (geography) was doing. This is interesting because it shows us that even on a natural level the universe can be separated into nearly watertight compartments. If it had not been so, everything would be much more complex. Science would have been much harder to discover, and indeed man might never have seen how things should be sorted out.

Aside from the importance of chemical energy as food, chemical energy has not to this day been as generally useful to us as electrical energy – a considerable heat tax has had to be paid on it whenever it has been transformed into other kinds of energy. The steam engine transformed chemical energy to energy of motion, but only by going first to heat and thereby incurring a huge tax. However, there has always been the hope of avoiding the intermediate step of going into heat, and perhaps this hope is now on the point of being realized through a device known as the chemical laser.

Nuclear energy is also converted first to heat and then to mechanical motion and electricity. Nuclear energy therefore suffers a high tax of 60 to 70 per cent, but this loss is less serious for nuclear energy than it has been for chemical energy because the amount of available nuclear energy is so enormous.

Energy can always be measured in the same unit, regardless of its form. We are probably all familiar with the commercial unit of electricity known as the kilowatt-hour (kWh). When you use an electric fire with a 1 kilowatt rating for a time of 1 hour you cause a flow of energy from electricity to heat, the amount of the flow being 1 kWh. You cause exactly the same flow by using a device with a 10 kilowatt rating for a time of $\frac{1}{10}$ hour (energy flow = power rating × time used). For the privilege of causing an energy flow of 1 kWh you pay in Britain (during daytime) about 2p, although you can cause the same flow at night for about 1p.

The practice grew up in the nineteenth century of measuring different forms of energy in different units; the energy of heat was in British thermal units (Btu), energy of motion in foot-pounds, and so on. This practice was bad, because it tended to prevent people from realizing that energy is the same stuff irrespective of its form. Only scientists really understood that this was so, and even scientists suffered from the need for curious conversion factors in order to relate the different units to each other. These conversion factors are hard to remember – off-hand I couldn't give you the conversion factor from foot-pounds to kilowatt-hours. So even scientists had to be always carrying a list of conversion factors around with them in their pockets. And for students the whole unhappy business has proved an abiding torment. So here I will use the same unit for every energy form. Since we all have acquaintance with the kilowatt-hour (those who pay household electricity bills certainly do) this will be the most sensible unit to use here.

Now how much of an energy flow (for energy in all its forms) do you need to maintain your personal way of life? How much for each day? In the United States about 250 kWh. This amounts to a flow of nearly 100 000 kWh for each year of life. Even if you are among the sturdiest of trenchermen only a small fraction of this 100 000 kWh is contained in the food you eat. Most of the energy we use (i.e. that we cause to flow from one form to another) goes in driving automobiles, heating houses and other buildings, smelting metals, and in the manufacture of goods. It is just because the food we eat forms only a small fraction of the total, a few per cent, that our society today differs from earlier human societies. In early times, and even in the days of Greece and Rome, food energy made up a much higher fraction of the total flow. There will be people who will tell you that we differ today from the past because of such things as law, or the United Nations, or the National Health Service, but pay no attention to such protestations. We differ because of energy flow.

Actually I have exaggerated a little the present general energy flow by giving the average per capita flow for the United States. The average for Britain is less, about 55 000 kWh per year, and this of course is the overriding reason why the standard of living is lower in Britain than it is in America. By doubling the British energy flow, our standard of living would rise inevitably towards the American level. In Britain, about 35 000 kWh per person per year goes on commercial activities; the remaining 20 000 kWh per year goes on personal use. (Burning 2 tons of coal or 500 gallons of heating oil consumes 25 000 kWh of chemical energy. Motoring 10 000 miles at 30 m.p.g. consumes about 15 000 kWh.)

Suppose now that we seek to raise the standard of living of everybody in the world to the American level. This would require an annual energy flow of about 100 000 kWh for each of the 4000 million people now living in the world. The total annual energy flow requirement would therefore be 400 million million (4.10^{14}) kWh.

There are three things to consider in relation to this last estimate, and the rest of this chapter will be concerned with discussing them. Does this estimate provide for the world's rising population? Would attempts to conserve the use of energy help very much? Is the seeming all-importance of energy affected by the possibility of exhausting reserves of metallic ores?

It is usually assumed that the world population will continue rising in the manner of Figure 3.1 until it reaches a total in the range from 10 000 to 15 000 million, where at last it will stabilize. The reasoning behind this point of view is entirely unclear to me. I can readily see how the world

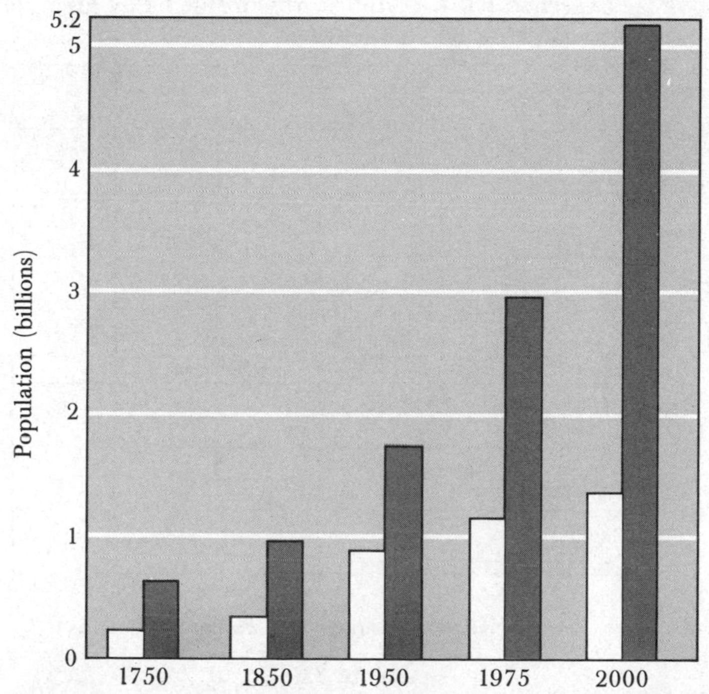

Figure 3.1 World populations, with light bars representing developed countries and solid bars representing developing countries.

population could be limited by disaster, but disaster is not the premise of the usual discussions. The issue is stabilization with continuing development. I find it much less difficult to see how stabilization might happen right now, before any further rise of population takes place, and I suspect that those who argue otherwise (see *Scientific American*, September 1976) do so only because they despair of checking the upward surge of Figure 3.1. Certainly there is no logical reason why rising populations should not be stopped,* and logic there had better be because world problems are far better solved now before reserves of coal, oil, and gas become threatened by immediate exhaustion. Since this book is concerned with solving world problems by logic, not

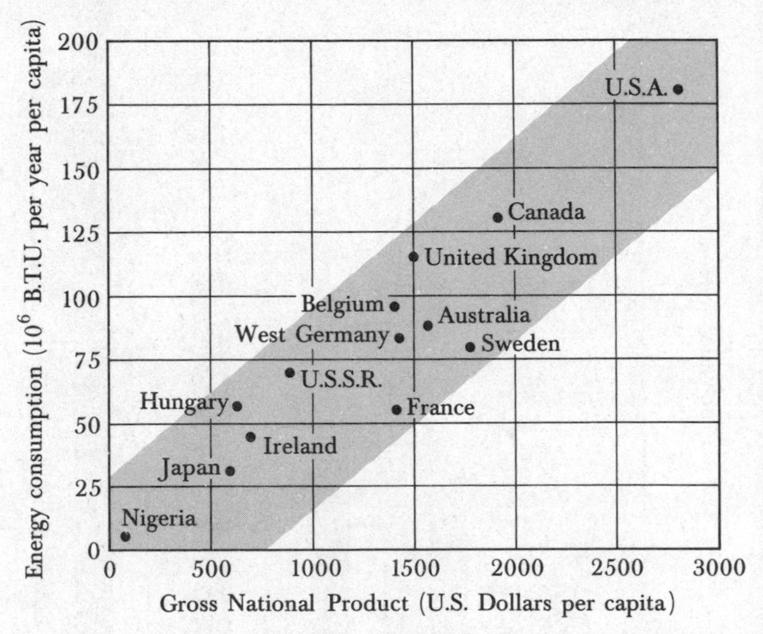

Figure 3.2

through disasters brought on by cultural flaws, I will take the position that the world population comes to be stabilized at its present level of 4000 million.

Turning now to conservation, it is not difficult to see how we all might save a little energy, but it is a mistake to think that conservation could go very far towards mitigating our

* This question is discussed in the final chapter of *Ten Faces of the Universe*, Heinemann, London 1977.

future need for energy. Conservation could moderate the need, but not by any great margin, for *if the margin were great, conservation would have happened already*.

To pursue the question of conservation in a little more detail, consider the plot shown in Figure 3.2. The dominant motif of this Figure is the general slant of the points representing different countries. This slant going from lower left to upper right demonstrates the basic relation between per capita energy flow and per capita Gross National Product (GNP). To the extent that the distribution of points forms a band rather than a precise straight line we may say there is scope for energy saving. But undoubtedly a part of the scattering of the points arises from causes other than energy-wastefulness.* The scope for conservation suggested by this Figure is no more than about 25 per cent. And any such gain from conservation will be largely compensated by the need for an increased per capita energy flow which must inevitably occur in the quite near future, within the next few decades.

In 1972 the 'Club of Rome' published a book entitled *The Limits to Growth* (Universe Books, New York). On page 56 there appears a table under the heading *Non-renewable Resources*. According to this table, the world's supply of metallic ores can last for only a few more decades. After raising this seemingly alarming problem, the book says little more than that we must mend our ways – we can't go on the way we've been going on. In so far as a solution is offered, it is 'conservation'. But no conservation policy could be total. Some metals corrode; others rub away into fine particles which could not be recovered; in many important usages at high temperatures, still others are partially vaporized. The most rigorous conservation policy could only extend lifetimes of a few decades into lifetimes of a few

* Part of the difference between France and the United Kingdom is due to a different balance between industry and agriculture.

centuries.* The solution to the apparent dilemma raised by the Club of Rome is so very straightforward that I would suppose it to be well-known, at any rate to every student of mining. The ore reserves used to compute the lifetimes of metal resources 'cut-off' at the concentrations which are currently judged to be 'economic'. Reserves become very much larger as ore concentrations are reduced, so that for low enough concentrations the metal content becomes essentially unlimited. The quantities of the metals in the ordinary rocks of the Earth's crust are enormous.† However, as the concentrations are lowered more energy is needed to crush the greater amount of rock that must inevitably be processed to yield the same quantities of metal. The correct interpretation of the table drawn up by the Club of Rome is simply that within a few decades a greater per capita energy flow will be needed in order to maintain our access to the same quantities of metals that we are using at present. As the required per capita flow increases, it will be a matter for rational decision as to how much energy should go for extracting new metal from the rocks and how much should go for conservation. Wherever one turns, it always comes down to energy flow.

Taking account of possible energy saving, and taking account of the problem of metal extraction, I think it reasonable to work in terms of a desirable annual per capita energy flow of 150 000 kWh, so that with the optimistic assumption of no further increase in the world population the total annual energy flow needed for the whole world is 600 million million (6.10^{14}) kWh.

Some people may think my per capita 150 000 kWh per year to be unrealistically high. But then I am probably

* And conservation would itself be expensive in energy usage. This is the reason why we don't have more conservation today.

† The quantities of many of the metals contained in the so-called manganese nodules, which lie on the floor of the deep ocean, are also very large.

being unrealistic too in assuming that the world population will stay fixed at 4000 million. Using a lower per capita energy, together with a higher future world population, leads to much the same total energy requirement. In a recent article, H. E. Goeller and A. M. Weinberg (*Science*, 1976, *191*, 683) arrive at a world energy requirement of 6.5×10^{14} kWh per year. These authors assume a per capita energy *average* of 65 000 kWh per year for a world population of 10 000 million, with about 130 000 kWh per year being used in developed countries. However one calculates the requirement, the needed total for a *politically stable world* cannot be much less than 6.10^{14} kWh per year.

4

Energy Availability: Non-nuclear Sources

There can be no disagreement with the statement that world reserves of coal, oil, and gas can provide an adequate energy source for only a limited future. Estimates of the lifetimes of these reserves evidently depend on consumption rates. Estimates will obviously be higher if world consumption is restricted to the present-day 10^{14} kWh per year, rather than expanded to 6.10^{14} kWh per year. The dilemma is that for consumption to remain at 10^{14} kWh per year the world's poor must remain poor. Yet at the desirable level of 6.10^{14} kWh per year lifetimes for coal, oil, and gas become alarmingly short, as we shall see later in this chapter.

Nor can it be contested that most of the world's population, presently 4000 million, will die in a disastrous catastrophe should an adequate energy source not have been developed by the time that reserves of coal, oil, and gas become exhausted.

Nor can there be any serious debate over the statement that the only alternative energy source *presently known to be technically viable* is energy from the nuclear fission of uranium or thorium.

The reader may well wonder, if these things are all true (and they are), why a strong campaign exists to stop nuclear energy from being developed. As the argument unfolds in this and in the remaining chapters the problem for the reader is to arrive at a plausible answer to this question. I have given my own answer in Chapter 1.

When those who speak strongly against nuclear energy are asked to explain what new energy source they themselves would advocate, the usual answer is 'solar energy'. When asked to provide precision engineering details of how an energy flow in excess of 10^{14} kWh per year is to be obtained from 'solar energy' they blandly admit to having none. They say that sometime in the future the details will become available, and they then go on to demand that someone should take the trouble to prove them wrong, not just in regard to a specific proposal but for every idea which occurs to them. Of course, it is very hard to prove that some startling new technology will not become available in the future, but a reasonable person would not gamble on it, particularly in view of the extreme gravity of the situation that would occur if the gamble failed. A reasonable person would push ahead as an insurance with what we already know to be viable. Then if something better becomes available in the future, well and good, belt and braces.

Allowing for the reflection by clouds of sunlight back into space, the flow of solar energy into the Earth's biosphere is about 10^{18} kWh per year, more than a thousand times greater than our needed energy flow. The flood of solar energy is so enormous compared with our need that *somehow*, one tends to feel, it should be possible to direct an adequate trickle of it into forms that would power our comparatively puny human industries. The case for solar energy really stops here. It has no follow-up. The lack of follow-up is worrying to most scientists and engineers, but less so to those who are not interested in details. The tendency of non-scientists is to feel that *somehow* it will be done. 'They' will find out the 'somehow'. It stands to commonsense doesn't it?

But does it really stand to commonsense? Plants and animals have been in fierce competition for sunlight over a very long time span, more than 1500 million years. By now they have become experts at using every last shaft of sunlight. Biological systems have evolved with a subtlety for grabbing sunlight that we ourselves could hardly match if we were to think for a thousand years about it.

'Ah,' it will be said, 'but *we* are not thinking about biological systems. *We* are thinking about artificial man-made systems.'

So let us take a look at this statement.

During the past two decades a technique based upon what is known as the photoelectric effect (the ejection by light of electrons from a metal surface) has been developed so that solar energy may be used to help power the radio transmissions of space vehicles and satellites. The device is usually referred to as a 'solar cell'. The recent edition of the *Encyclopaedia Britannica* (Vol. 2, 769) has this to say:

> 'Little use has been made of (solar) cells in terrestrial operation, owing . . . to the poor efficiency of the cells, and their high cost.'

For definitions, I will consider a large collecting area, subdivided into 'cells' each with an area of 10 square centimetres, and I will suppose that future technology increases the cell efficiency from the present 12 per cent to 20 per cent and brings down the price (with interconnections) to only 1p per square centimetre. To collect 0.06 per cent of all the sunlight not reflected by cloud (i.e. about 6.10^{14} kWh per year) our cells would be required to cover about 1 per cent of all the land surface of the Earth, an area of about 1 million square kilometres, an area comparable in size to Western Europe. With each cell 10 square centimetres in area there would need to be 1 000 000 000 000 000 of them, and they would cost (even at our low price) £100 000 000 million. By now the imagination begins to reel. But we still have to collect all the minute quantities of electricity emerging all the time from each of our 10^{15} cells. How is this to be done with good efficiency? It isn't, of course, and only irresponsible people would let the lives of 4000 million people turn on such an expensive, untried idea.

We might think of focusing sunlight with mirrors, perhaps using the focused sunlight to boil water. But even if each mirror covered 10 000 square metres (larger than the radio-telescope at Jodrell Bank), and even if the systems were 100 per cent efficient, there would still need to be 20 million water boilers. Steam would need to be piped from each of them to central power stations, without loss of efficiency, and an area of several hundred thousand square kilometres would need to be covered by mirrors. This project too is nonsense. Small-scale solar projects for the home may make sense, if you can afford one of them, but I have not yet seen any large-scale sunlight-collection project that would be likely in the foreseeable future to deliver as much energy in its use as it consumed in its manufacture. This I believe to be also the case for projects that seek to use sunlight collecting areas out in space.* It would clearly be unwise, to the point of foolishness, to pin our whole future on untried, remote schemes.

Far better to let plants do the collection of sunlight for us. Think of the plodding human, taking care to position and wire up his 10^{15} cells, and then think of biology scattering a profusion of seeds in the wind. The marvellous subtlety of biology is that it broadcasts its solar energy collectors in embryo form, for seeds are not yet energy collectors. They *grow* into energy collectors. Another marvellous thing is that the chemical energy produced from sunlight *accumulates* in a plant. There is no requirement for it to be tapped continuously. In ground plants, the chemical energy

* Some scientists would disagree with this statement, but the very large research and development funds which they admit to being necessary show that such projects are still quite visionary.

accumulates for several months; in trees, it accumulates for many years. It is just because the chemical energy accumulates for much longer in trees than it does in ground plants that we prefer to burn wood as a fuel rather than, say, straw. But splendid as this may be, dependence for energy on trees

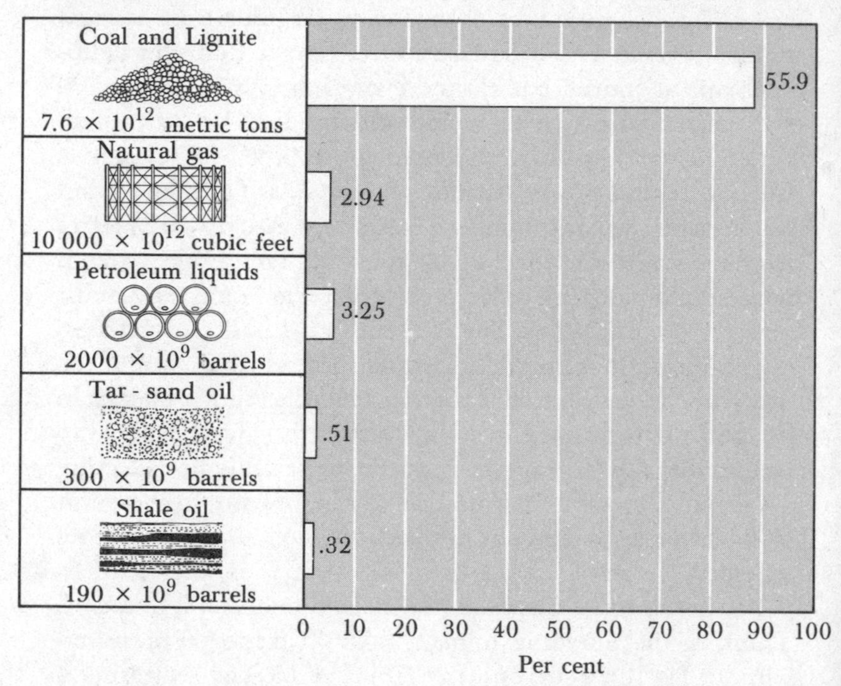

Figure 4.1 World reserves of coal, oil, and gas (after M. King Hubbert, *Energy and Power*, W. H. Freeman, San Francisco, 1971)

would take us back to just the situation which existed before the incidence of modern industry. We should simply be back in the days of the old charcoal burner. There is no route here either.*

The energy stored in straw, wood, coal, oil, and gas is all chemical; they are all for the most part of the same nature, all hydrocarbons. The difference between these fuels lies in

* See note at the end of this chapter for further comment.

the time span of the storage; months for straw, tens of years for wood, and still longer effective storage times (because of compression which occurred underground) for coal, gas, and oil. It was the concentration of the energy stored in coal which provided the springboard for the development of modern industry. *Concentration of energy* is as critical as the amount of the energy, and it is energy concentration which is so difficult to achieve from diffuse sources like sunlight or like ocean currents and tides and wind.

At this point it will be as well to take a look at world reserves of coal, oil, and gas. These are shown in Figure 4.1 where the numbers are energy totals with 10^{15} kWh as unit. To supply the energy requirement of 6.10^{14} kWh per year arrived at in the preceding chapter, all the world's initial

Table 4.1 *Lifetime of world reserves of hydrocarbons for a depletion rate of 6.10^{14} kWh per year*

Oil		*Lifetime (years)*
Proved reserves		1.79
Probable reserves		0.34
Still undiscovered reserves (estimated)		2.51
	Total	4.64
Gas		
Proved reserves		1.01
Probable reserves		0.18
Still undiscovered reserves (estimated)		2.21
	Total	3.40
Coal		
Proved recoverable		5.9
50% of probably recoverable		49.4
	Total	55.3

Based on data from U.K. Department of Energy, presented at the National Energy Conference, June 1976

reserves of oil and gas would last for a mere 10 years, while reserves of coal and lignite would last for about 80 years. Present-day lifetimes are of course less, because in each case a portion of the initial supply has already been used up. Table 4.1 shows the current (1976) estimates of reserves given by the U.K. Department of Energy. The lifetimes are calculated in each case for an energy depletion rate of 6.10^{14} kWh per year.

Only 50 per cent of the energy from probably recoverable coal has been included in Table 4.1 because much of the world's coal exists in thin seams and a considerable energy input is required to recover it. The energy 'profit' from the coal is therefore less than the energy content of the coal itself. Some people might think a 50 per cent recovery of actual energy to be pessimistic, but others think it optimistic, and at all events the precise value to be attached to the net energy recovery could not change the fact that all lifetimes calculated for the hydrocarbons are alarmingly short.

The lifetimes are extended sixfold if the present-day world energy consumption of 10^{14} kWh is used in the estimates instead of 6.10^{14} kWh. But 10^{14} kWh makes no provision for relieving the poverty of most of the people living on the Earth, nor for any increase that may occur in the world population. Manifestly there will be an insistent demand to increase the present-day energy consumption, a demand that will bring the exhaustion of the hydrocarbons to within the lifetimes of one's grandchildren unless a new and adequate energy source is forthcoming. Without such an energy source, the world will be wracked by political troubles whose intensity will surely increase as the moment of eventual exhaustion of the hydrocarbons approaches.

An enterprising engineer living close-by a natural hot spring might think of piping off hot water for his home. And a whole village or even a small town might sensibly seek to tap a local geothermal resource. But there is no possibility of geothermal heat making a significant contribution to the energy need of the whole world since the heat emerging from the Earth *taken over the entire land surface* is less than 10^{14} kWh per year (our target is 6.10^{14} kWh per year). The fuss which has been made in public debate over geothermal energy should therefore add to a forming suspicion that the real motive of those making the fuss has not been to propose a sensible new kind of global energy resource.

Wind and water power were the main energy sources of post-medieval industry, and they have played highly respected roles in the history of technology. But the fact that these energy sources were already proving inadequate in the seventeenth and eighteenth centuries does not encourage the view that they are likely to stage a profound come-back in the twenty-first century. Engineering systems for these energy sources are well-developed, probably with little that is radically new to be added.

Although to the modern eye windmills have a picturesque image, it was not always so. Writing in *A History of Technology* (Oxford University Press, 1958, Vol. IV, page 156) Professor R. F. Forbes remarks:

> 'The introduction of the windmill as a prime mover did not proceed smoothly. The craftsmen's guilds of Holland protested against them in 1581, claiming they would throw many craftsmen out of work. A mechanical saw worked by a windmill was built in 1766 at Limehouse, east of London, but was destroyed by a riotous mob two years later.'

Professor Forbes gives a table for the energy output of a windmill fitted with modern sails of span 100 feet. The annual energy output expressed in kWh is shown in Table 4.2 for various assumed wind speeds. Windspeeds as high as 9 metres per second occur by no means all the time, probably not more than one-third of the time. There would be energy

Table 4.2

Wind speed (metres per second)	*Annual energy output (10^5 kWh)*
3	0.61
4	1.40
5	2.71
6	4.38
7	6.92
8	10.2
9 and over	14.6

Based on data given by R. J. Forbes, A History of Technology, *Vol. IV, page 156 (Oxford: Oxford University Press, 1958)*

losses in the collection of power from the very many mills that would be needed to maintain a major industrial complex – losses from the vast number of electrical channels coming from the separate dynamos attached to each windmill – and there would be losses from the vast number of electric storage batteries that would be needed to buffer the electricity supply in order to maintain power on windless days. For an effective annual per capita energy production of 150 000 kWh it is therefore clear that one such windmill would be needed for every two or three persons. The number of mills required by Britain alone would thus be about 20 million. Taking each one to stand on a square of ground with side equal to twice the sail span, the area occupied by 20 million windmills would cover more than half the area of all England. While increasing the sail span above 100 feet would increase the output from each mill (according to the area swept out by the sails) the ground space needed by each mill would increase correspondingly and nothing much would be saved in the total area required to accommodate all the mills. When in full operation such an ensemble of mills would make an appalling roar, and the number of serious accidents they would cause would probably run into hundreds of thousands each year.

Suggestions for using tides and waves as a main energy

source are also impracticable. It has been calculated that the wave-energy incident on a coastal boom of length 1 mile might amount to an annual total of 10^9 kWh. Although the collection and concentration of such a diffuse energy source would be an engineer's nightmare, if we assume the energy to be collected with 100 per cent efficiency in some manner, 10^9 kWh per year would provide 150 000 kWh per year for each of about 7000 persons. Britain alone would require a boom length of about 8000 miles, while an energy total of 6.10^{14} kWh per year for the whole world would require a fantastic boom length of 600 000 miles, a boom length two-and-a-half times the distance from the Earth to the Moon.

People who recommend energy sources of this kind usually aim for much lower per capita yields, usually lower than 50 000 kWh per year. Quite apart from the social resistance there would be in the western democracies to the resulting fall in the standard of living, such a reversion to a hair-shirt economy would lead to short-term disaster since hair-shirt economies cannot support difficult, cumbersome technologies. It is an important matter of principle that access to energy should not consume too much of the working capacity of society otherwise a descending spiral would soon start up with a recession of the economy reducing the energy supply, which in turn would deepen the recession, reducing the energy supply still further. A weak economy that sought to collect and to concentrate diffuse energy from a multitude of windmills or from thousands of miles of coastal boom, *an economy without appeal to other energy sources*, would soon collapse under the weight of its ponderous engineering systems.

It is true that post-medieval society made use of wind and water in a reasonably stable way, but post-medieval society only supported a much smaller population which also had access to wood as an energy source; and post-medieval society supported only a cottage-industry standard of living. This would be the way of it for a wind-and-water economy in the future. The population would need to fall drastically from its present level, and there would be an inevitable return to a much lower living standard.

Another better form of water power is also worth considering in a little detail because this better form of water power brings out two further important points. The first is that whenever a genuine opportunity of access to a new energy source comes up, people always seize it. Environmentalists are always accusing governments of being derelict in not paying sufficient attention to ideas for using solar energy, and they are always saying that the expenditure of a few million pounds on this or that project would soon transform the energy problem. The case of water power shows that neither governments nor people generally ignore sensible new possibilities. The basic idea of water power has not changed since classical times – to transform the energy of position of water, as it flows downhill, into the energy of motion of a machine (and thence in modern times into electrical energy). What *has* changed in modern times is that now we have far better access into high mountain regions where large amounts of precipitation occur and this improved access has been effectively used in our modern hydroelectric projects. Today hydro-power delivers about 10^{13} kWh per year, much less than our needed total but still an entirely respectable total, well worth achieving – for hydro-power really does deliver more energy in its operation than it consumes in its manufacture. There is still scope for sensible new hydroelectric projects, particularly in Africa and South America, but there is no possibility of solving the world's main energy problem in this way.

The second point about hydro-power is that it shows in a graphic manner just where the trouble lies in other sunlight-collection schemes – it brings home to us the sheer scale of the problem of solar energy. Sunlight absorbed by the ocean causes evaporation of water vapour into the atmosphere.

Some of the water vapour is carried over the land by winds and some of it condenses as clouds, particularly over the mountains. Through rain and snowfalls water comes to be deposited on high ground. The rain, and the snow when it melts, then runs downhill, turning energy of position into energy of motion. At first the water runs in a multitude of tiny trickles. Then the trickles aggregate into small streams, the small streams aggregate into larger ones, until eventually a fast-moving torrent is formed. And many torrents join at last into a formidable cascading river. This aggregation of water forms the natural collecting system for hydro-power. We ourselves are not required to perform the collection, natural drainage channels in the land surface do the collection for us, and this is precisely why hydro-power is successful. If man had been required to do the sculpting of the land surface himself hydro-power would not have been successful, for the task of shaping the land would have been too energy-consuming; and even with sufficient energy, forming the mountains themselves would surely have been beyond our capacity. The example of hydro-power gives our common-sense something firm with which it can get to grips. It gives us a stable platform from which we can see that so many schemes for the collection of solar energy scarcely need weighing in the balance. They are so obviously wanting.

* (ref. page 32) In a recent article (*Science*, 1977, **196**, 613) B. Bolin has given biomasses that are accumulated annually for the whole Earth in forests, oceans, cultivated crops, swamps, and marginal lands. If one takes account of the extra energy that would be necessary to collect all this material, not even burning it all would meet an energy requirement of 6.10^{14} kWh per year.

5
Energy Availability: Nuclear Sources

Turning back to Figure 2.1, it is interesting to notice that atoms of the greatest mass are enormously less abundant in the Sun than the lightest atoms. Thorium and uranium are the two heaviest atoms shown, and they are about 1 million million (10^{12}) times less abundant in the Sun than hydrogen. Over the past generation astronomers have learned that all the various kinds of atom that were present inside the gas cloud from which our solar system condensed (excepting the low abundance group Li, Be, B) were produced from hydrogen and helium by nuclear processes that occurred inside previously existing stars. Carbon, oxygen, and neon, are produced from helium; and nitrogen is produced from carbon, oxygen, and hydrogen. Sodium, magnesium, aluminium, and silicon are also produced from carbon and oxygen, while phosphorus, sulphur, chlorine, argon, potassium, calcium, scandium, titanium, vanadium, chromium, manganese, iron, cobalt, and nickel are produced from silicon and magnesium. All these processes take us to elements of atomic mass around 60.

Is further nuclear development inside stars possible to still heavier elements like platinum, thorium, and uranium? Except in the sense of minor by-products the answer to this question is, no. This answer, which we know to be right from laboratory studies of nuclear reactions, sets the stars a great problem. For just as a man-made nuclear reactor supplies us with an energy flow, so the nuclear reactions inside stars supply them with energy flows. And stars have a continuing need for energy to make good what they are losing as they shine out into space. But the known fact that stellar nuclear reactions reach an end, an end in which atoms of magnesium and silicon serve as a fuel for producing heavier atoms like iron and nickel, means that the nuclear energy supply must dry up. What are the stars to do then? This question raises the problem of the final fate of a star, of its ultimate graveyard.

The problem is a complex and fascinating one for astronomers. The road may lead in some cases to those mysterious objects known as *black holes*. In other cases, the road leads to explosion which disintegrates much, if not all, of the star. The violence of these explosions, *supernovae* as they are called, involves a sudden and final release of energy equivalent to 100 million million million million (10^{26}) hydrogen bombs all exploding together. Besides being very spectacular, these explosions broadcast far and wide the many kinds of atom which have been produced within them, the many kinds of atom which make up the materials of our everyday world.

The explosions of stars set up the cyclic relation of Figure 5.1. Through this relation we are now in a position to understand how the rich variety of 90 kinds of atom arrived here on the Earth. They were produced in stars which completed their lives before the Sun and our planetary system were formed. Our galaxy, the Milky Way, is about 12 000 million years old, nearly three times the age of the Sun and Earth. The preceding generation(s) of stars broadcast their synthesized materials throughout our galaxy by means of explosions. And so it was that the many kinds of atom came to be present within the interstellar cloud of gas from which our Sun and planets at last were born.

We noticed already in Chapter 2 that the Earth formed in such a way that there were large deficits of the light elements, leading to our present lack of hydrocarbons. And we also noticed that the Earth formed in such a way that the abundances of uranium and thorium were markedly increased throughout an outer skin of rock. Like all the heavy kinds of atom, uranium and thorium were produced through minor by-product reactions inside stars, and they were broadcast through space by explosions.

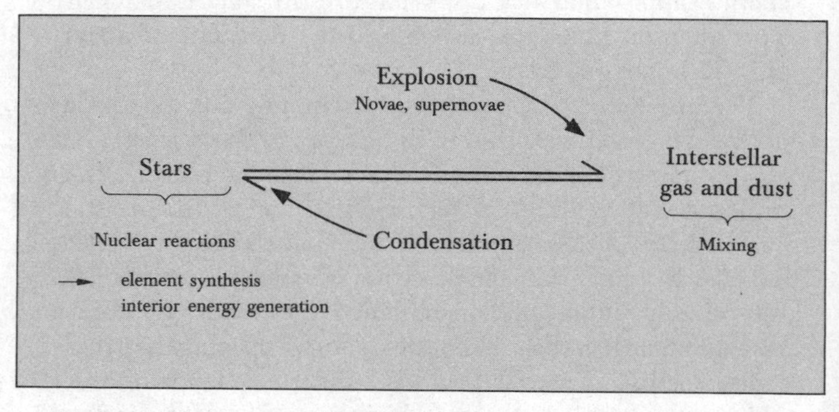

Figure 5.1

The reason why very heavy atoms are only produced in reactions of a by-product character is that no energy comes from their synthesis. In fact, just the reverse. Energy has to be supplied in order to produce atoms with masses greater than about 60. Only in very rare situations does this happen because stars, like humans, are seeking to obtain energy, not supply it. But occasionally, in somewhat fluke conditions, these by-products are made, and it is because energy happened to be supplied to produce them that today we are in some degree able to draw on this energy, as we do when uranium or thorium is used as the fuel of a man-made nuclear reactor. We are then releasing energy that was stored in the stars many thousands of millions of years ago.

In principle it would be possible to derive energy from the break-up of any one of many forms of heavy atom – lead, mercury, barium, to give three examples – indeed from any atom with mass somewhat greater than 60. For reasons of nuclear physics concerning the method used to induce break-up, the technology for doing so exists only for the use of uranium and thorium. In concerning ourselves with practical issues, the discussion must therefore be confined to the fissional break-up of uranium and thorium only, though we have recognized here that wider possibilities may exist.

If the nucleus of an iron atom, shall we say, is thought of as being an inch in diameter, then the lightweight electron cloud around it would have a diameter of about a quarter of a mile. The nucleus of iron is an aggregation of two kinds of particle, neutrons and protons; so are the nuclei of uranium and thorium. The thorium we find here on Earth has a nucleus containing 90 protons and 142 neutrons. We can denote this by writing Th (90p, 142n). Naturally occurring uranium comes in two forms, however, known as *isotopes*. In a similar notation, the two isotopes of uranium can be written as U (92p, 143n) and U (92p, 146n), from which you see that the two isotopes have the same number of protons but different numbers of neutrons.

The two isotopes of uranium are more often written as U-235 and U-238, but if you write them in these forms it is necessary to remember that the uranium atom is characterized by having 92 protons. Then you can find the numbers of neutrons by simple subtraction, $235 - 92 = 143$ for U-235, and $238 - 92 = 146$ for U-238. Similarly, it is more usual to write Th-232 for naturally occurring thorium; here again you must remember that thorium is characterized by having 90 protons, so that the neutron number follows by subtraction, $232 - 90 = 142$. From here on, I shall write Th-232, U-235, and U-238.

It is comparatively easy and convenient to break-up U-235 but not nearly as convenient to break-up U-238 or Th-232, again for reasons of technical nuclear physics. And the rub of it is that the easy case, U-235, constitutes only 0.715 per cent of naturally occurring uranium. Totally separating the small fraction of U-235 from the bulk U-238 is still a difficult and expensive process, so much so that it is not done for civil energy-producing purposes. What is often done is to concentrate U-235, up to about 3 per cent of the U-238. Such 'enriched' uranium is used in the fuel cells of most forms of civil reactor, but not in what I believe to be the best form of nuclear power reactor yet designed, the CANDU reactor. (Canada-can-do-it, a justifiable pat on the back for the Dominion.)

The technique for breaking up any one of Th-232, U-235, U-238 is to introduce an additional neutron into the nucleus. For Th-232 and U-238 the additional neutron must be fast-moving, but for U-235 the additional neutron can either move fast or slow. Slow-moving neutrons are less awkward to handle than fast-moving ones, and this is why reactors based on U-235 are much less difficult to design than the so-called 'fast reactors' using U-238. When U-235 breaks up (into two more or less comparable pieces) on the average rather more than 2 neutrons become 'free'. That is to say, the neutrons escape out of the broken-up pieces. These free neutrons are fast-moving at first, but by attention to the geometrical design of the reactor and by introducing a suitable agent known as a 'moderator', which is sometimes ordinary water, sometimes heavy water, and sometimes carbon blocks, the neutrons can be slowed. After slowing, rather less than a half of them are used to break up more of the U-235.

Since rather more than 2 neutrons per fission were produced in the first break-up, using more than a half of these first neutrons to break up more of the U-235 would set going an amplifying cycle. The fission process would 'runaway'. Now there is a wonderfully lucky thing about the way that neutrons happen to come free during break-up which can be used to stop runaway from occurring – a fraction of them come free only after a rather long delay. This delay gives ample time for either control rods which absorb neutrons very strongly to be moved in or out of a reactor or for a neutron reflector to be changed should the cycle not be in precise adjustment. The strong neutron absorption by the control rods then stops the runaway. It is this lucky break which makes it relatively easy to operate nuclear power plants in a stable and safe way.

Because only rather less than a half of the free neutrons is needed to keep the fission cycle going, the other half is 'spare', and the question arises as to whether this spare half can be put to use. Because the moderator has slowed the free neutrons they do not cause the bulk U-238 (which constitutes most of the material of the fuel cells) to undergo fission. But U-238 can absorb a neutron without breaking up. The resulting U-239 does undergo a nuclear reaction, however, but a far milder one than break-up itself; the U-239 spits out an electron and changes to Np-239, which again spits out an electron and changes to Pu-239. Here Np stands for an atom of neptunium, and Pu for an atom of plutonium. The absorption of a slow-moving neutron by U-238 therefore changes the chemical nature of the atom itself, from uranium to plutonium. A new kind of atom is born, one which does not occur naturally on the Earth (except in an exceedingly minute amount).

Another remarkable thing now emerges. The break-up properties of Pu-239 are quite similar to those of U-235; Pu-239 can be made to undergo fission by either fast or slow neutrons. So Pu-239 will do essentially just as well as U-235 to maintain the fission cycle. Particularly, it will yield rather more than 2 free neutrons per break-up, one of which we can use to maintain the cycle itself, and the other we can use to make still more Pu-239 from U-238. In principle we shall have made as much new fuel out of U-238 as we have used of the U-235. And when all the U-235 has been broken up we can simply go on and on with the Pu-239 until all the U-238 has also been broken up. From the point of view of fuel economy we shall then have made effective use of about a hundred times more of the uranium than we might have expected to do in the first place. This process is known as *breeding*.

The importance of breeding for the economy of nuclear power is that it offers the hope of obtaining the break-up energy of all the uranium instead of only the 0.715 per cent that is U-235. But breeding will fail if atoms other than U-238 succeed in grabbing the spare neutrons. By attention to the design of the reactor it is possible to prevent the moderator and the fuel cell containers from being too much of a nuisance in this respect. However, within the fuel cells themselves pieces from the break-up of the uranium will accumulate and among the accumulating pieces are some avid neutron-grabbers, known technically as *poisons*. Still, by cleaning out the poisons from time to time – *reprocessing* as it is called – this difficulty can be overcome. Then a third, more insidious, problem remains. In only about 70 per cent of cases does Pu-239 undergo break-up when it adds a slow-moving neutron. In the remaining 30 per cent of cases, the isotope Pu-240 is formed. Thus Pu-239 itself acts like a poison, and so does Pu-240.

The serious self-poisoning effects of Pu-239 and Pu-240 have so far limited the 'burn-up' of U-238 to about 1 per cent (in reactors operating commercially). This unfortunately is not much better than the 0.715 per cent of the U-235. Two ways for obtaining a marked improvement of this still poor fuel economy are known, however. The two ways are markedly different in concept and method, so that an explicit choice between them involves a decisive commitment to a particular form of technology. I will discuss the better-known system first, although this better-known system does not happen to be the choice that I would make myself.

A high burn-up percentage for the U-238 can be achieved by removing the moderator. The initially fast-moving free neutrons will remain fast-moving, and they will be capable of inducing break-up of even U-238. A fission cycle based on U-238 then becomes possible. By stopping such a 'fast' reactor from time to time much greater quantities of Pu-239 can be extracted chemically from the fuel cells than is possible for the 'slow' reactors that use a moderator.

Although these 'fast breeder reactors' have the great advantage of high fuel economy, they have the disadvantage of being harder to control because they work at much larger energy concentrations than the slow reactors. Indeed the control problems are such that (at any rate for the foreseeable future) the typical nuclear power plant could not be of this kind. What might be done, however, would be to supply a typical nuclear power plant, which would remain of the 'slow' kind, with Pu-239 produced in a special fast reactor at which exceptional precautions were taken, for example by having a larger and more skilled technical staff on hand than would be necessary for a slow reactor.

Fast breeders have given anti-nuclear environmentalists a huge stick to beat the nuclear dog with. Fast breeders have been in successful operation on an experimental scale for more than a decade, but no large-scale breeder suited to practical power generation has yet come into operation (although there are plans for them to do so). Experience with experimental models has thrown up technical problems, problems which have always been openly discussed. The need for these problems to be solved has given organizers of the antinuclear campaign the opportunity to agitate for all development towards nuclear energy to be delayed until *after* these admitted problems with fast breeders have been demonstrably solved. Such a delay would probably be long enough to permit the Soviet Union easily to win the world struggle for energy.

It has also been argued that the shipping of Pu-239 from a few central stockpiles out to many power plants would open the way to plutonium theft. Especially would this be true, it is argued, if Pu-239 were shipped up and down the world on an international scale. In an article which I shall mention more fully in the next chapter, Bernard L. Cohen* has the following to say on this issue:

> Theft of Plutonium
>
> Widespread concern has been expressed that plutonium may be stolen from the nuclear energy industry by terrorists for fabrication of nuclear bombs. While this threat cannot be quantified, it cannot be ignored in assessing the environmental impacts of nuclear power.
>
> The principal protection against this threat is to prevent plutonium from becoming available to prospective terrorists. The method for keeping track of it since the 1940s has been to weigh all plutonium entering and leaving a facility, but errors in weighing leave substantial room for undetectable losses. It is not impossible that enough plutonium has already been diverted to make many bombs. On the other hand, there is no evidence that plutonium has ever been stolen in quantities as large as one gram.
>
> The A.E.C. has long conducted programs for improving security, but in 1973, T. B. Taylor, a former bomb designer, became disenchanted with the slow progress in these programs and 'blew the story open' in a remarkable series of articles in the New Yorker magazine and a book (34) in which he gave information on how to make

* Reprinted by permission of *American Scientist*, journal of Sigma Xi, The Scientific Research Society of America.

nuclear bombs. As a result of the publicity he received, plutonium safeguard procedures were greatly tightened and new regulations are constantly being added.

Clearly, the issue of terrorism should have been considered in the 1950s, before the nuclear industry began. It seems almost irresponsible to raise the problem after so much money and effort has been expended on the industry and we need its product so badly. Nevertheless, the threat of terrorism has continued to escalate in importance, and it is now one of the principal points of contention in the nuclear power controversy.

A number of factors should be considered in evaluating the risk of terrorism. First, stealing plutonium would be very difficult and dangerous under present safeguards. Taylor has estimated (35) that a group of thieves would have much less than a 50% chance of escaping with their lives. Fabrication of a bomb from stolen plutonium would also be very difficult, expensive, time-consuming, and dangerous. Estimates vary considerably, but a rough median of the opinions of experts indicates that it would require three people highly skilled in different technical areas a few months and perhaps $50,000 worth of equipment to develop a bomb with a 70% chance of doing extensive damage, and that the people involved would have a 30% chance of being killed in the effort.

Terrorist bombs would be 'block-busters', not 'city-destroyers'. Taylor's principal scenario (34) is that an explosion of this sort could blow up the World Trade Centre in New York, killing the 50,000 people that building can contain. Of course, there are many much easier ways to kill as many people (e.g. introducing a poison gas into the ventillation system of the World Trade Centre).

In the past, terrorists have had options for killing thousands of people, but they have almost never killed more than a few dozen. And, of course, plutonium, and highly enriched uranium that would be much more suitable for bombs, could be obtained from sources which have no connection with nuclear electric power.*

In this book I purposely avoid these problems associated with fast breeders, not because of serious doubt that they can be solved, but because the technical facts of the case seem to me to point clearly towards the second of the two ways in which an improved fuel economy can be obtained. For the technically equipped reader the data of Table 5.1 bring out

Table 5.1 *Thermal Neutron Cross-sections (barns)*

Atom	*For neutron capture*	*For fission*
Th-232	7.3 (wanted for breeding)	
U-233	53 (unwanted)	525 (wanted)
U-234	90 (unwanted)	
U-235	106 (unwanted)	577 (wanted)
U-238	2.7 (wanted for breeding)	
Pu-239	286 (unwanted)	742 (wanted)
Pu-240	290 (unwanted)	

the crucial point that *breeding from Th-232 in 'slow' reactors is much less subject to self-poisoning* than is breeding from U-238 (technically, because the unwanted neutron capture cross-sections for U-233 and U-234 are much smaller than the unwanted cross-sections for Pu-239 and Pu-240). Breeding from Th-232 is particularly adaptable to the CANDU-type of 'slow' reactor. With comparable amounts of uranium and thorium in the fuel bundles of this form of reactor spare neutrons are more likely to be added to Th-232 than to U-238.

The resulting isotope U-233 of uranium has fission

* In relation to this last remark, there is the possibility (outside the nuclear power industry) of producing highly enriched uranium by a laser technique. If this proves relatively simple, the whole plutonium argument used by environmentalists becomes worthless, because anyone making illicit nuclear bombs would then seek to use enriched uranium, not plutonium.

properties that are closely similar to U-235. Use of the CANDU system therefore avoids the still unsolved problems associated with fast breeders. And by keeping the U-233 mixed with U-238, the danger of illicit bomb-making can also be avoided (but see previous footnote).

Dr W. B. Lewis, the designer of the CANDU system, considers that a burn-up of 5 per cent of the fuel could currently be achieved by mixing thorium with uranium, and he thinks that the burn-up (the fraction of the fuel to undergo break-up) could eventually be increased very significantly, perhaps to as much as 50 per cent. Nuclear physicists in the United States with whom I have discussed this question have all agreed that a 5 per cent burn-up should be attainable more or less immediately by the CANDU system. In the following discussion I shall therefore use this percentage, although future technical advances could well improve upon it. Explicitly, what I propose to show is that all our problems of energy availability can be solved by very safe reactors of the CANDU type.

The Pickering reactor built for Ontario Hydro was the first major CANDU-type reactor to be operated commercially. Although it is usual for such first-off-the-line projects to give teething troubles, Pickering ran well from the beginning. In 1975 Canadian dollars it delivers electrical energy at a cost of 0.703 cents per kWh, which is low compared to 2p in Britain (daytime) and low compared with U.S. charges which are now rising to about 2c per kWh. The cost of electrical energy has increased very rapidly in recent years because of the steep rise in the price of coal and oil, and because most of our power plants are fired by coal or oil. In the nineteen-eighties, when the hydrocarbons become more expensive still, the economic advantage of nuclear energy over hydrocarbon energy will become even more marked. All this is shown very clearly by the comparison of Table 5.2 (1m$=0.1 cent).

Table 5.2 *Ontario Hydro 2000 MWe generating station costs, March 1975 dollar values; maturity cost estimates for 80% net capacity factor*

Station	*Nuclear* *Pickering*	*Coal* *Lambton*	
Capital (m$/kWh)	4.60	1.70	1.70
Operating and maintenance	1.10	0.96	0.96
Fuelling	0.98	10.60[1]	13.52[2]
Heavy water upkeep	0.35		
Total unit energy cost	7.03	13.26	16.18

[1] Current coal inventory [2] New coal for inventory

From L. W. Woodhead and L. J. Ingolfsrud, European Nuclear Energy Conference, Paris, 21–25 April, 1975

I will suppose now that you would be willing to pay $\frac{1}{3}$p per kWh to cover the cost of uranium/thorium fuel. It is then easy to calculate how much we can afford to pay for a kilogram of fuel. And knowing what we can pay for fuel, we can make a further estimate of how much uranium/thorium can be found the world over at that price. Obviously if we can afford to pay a lot, more fuel will be available than if we can only afford to pay a little, because the higher the price the more it becomes profitable for mining companies to extract uranium and thorium from large reserves of low grade ore. Then knowing the amount of recoverable uranium and thorium the world over we can determine the length of time for which our reserve of nuclear fuel would be capable of delivering the required world energy flow of 6.10^{14} kWh per year.

Before proceeding to this calculation it is worth noticing that the strict requirement for estimating the availability of uranium/thorium fuel is that more energy should be derivable from the fuel than was expended in its acquisition. However, we saw already in Figure 3.2 that a good general correspondence exists between energy expenditure and the monetary value we attach to commercial activity. Over the broad spectrum of industrial activity, as one has in the manufacture and operation of mining equipment, or as one would have in the chemical extraction of uranium from seawater, energy expenditure and monetary costs are so well-correlated that money values can be used as a measure of energy values (provided only that the margin of the calculation is not tight – it certainly will not be tight here).

At a 5 per cent burn-up 1 kilogram of uranium/thorium fuel would produce about a million kilowatt-hours of heat energy. Conversion of this heat to electricity at an efficiency of 30 per cent gives 300 000 kWh of electrical energy. Given ⅓p per kWh of electricity as our payment towards the cost of fuel, we could evidently afford to pay £1000 per kilogram of uranium/thorium.

Suppose then that we offer to buy uranium and thorium at this price. How much can we expect the world's mining companies to supply us with? A very large amount certainly. More than all the uranium in seawater, which amounts to about 3000 million tons, because the uranium in seawater could be extracted at a cost of about £200 per kilogram. It would also be possible to extract uranium from low grade ores on the land. The amount of these ores is known to be very large indeed, but the exact amount is not known because nobody has yet done the necessary geological survey (to cover the case of a uranium or thorium price as high as £1000 per kilogram). Very conservatively, I would suppose that 10 000 million tons of uranium would be forthcoming, and since thorium has an average concentration in rocks that is about three times higher than the uranium concentration it would be even more conservative to take 10 000 million tons as the amount of thorium which could be similarly obtained. This gives a total of 20 000 million tons for the amount of uranium/thorium to be obtained by paying ⅓p per kWh of electricity.

The rest of the calculation is quickly done. 20 000 million metric tons is 20 million million kilograms. With each kilogram giving 1 million kWh of energy, the total thermal yield would be 20 million million million (2.10^{19}) kWh. Drawing on this vast reserve at a rate of 6.10^{14} kWh per year, the reserve would last for about 30 000 years.

It is inconceivable that technical developments over tens of millennia would not raise the 5 per cent burn-up assumed in the above calculation. A 50 per cent burn-up (which I would expect to be accessible with thorium fuel enriched by, say, one per cent U-235) would itself increase the lifetime to several hundred thousand years. It would also greatly increase the price we could afford to pay for thorium. At ⅓p per kWh and at a 50 per cent burn-up, we could afford £10 000 for each kilogram of fuel. The amount available at such a price would be truly enormous, extending still farther the lifetime of our nuclear resources, surely from hundreds of thousands of years to many millions of years.

The problem of long-term access to an adequate energy source is therefore solved. It is solved for many thousands of years without needing any really major projections beyond what has been done already – the technology is either accessible now or accessible within a comparatively short-term future, a decade or two at most. It is solved without fast breeders. It is solved without the outrage to commonsense which proposals for solar energy and other forms of non-nuclear energy always involve. There is no requirement for Britain to have millions of reactors, as there would be for windmills, or for having thousands of miles of coastal boom as there would be for wave-energy. About 200 reactors for the whole of Britain would suffice.

Since this chapter is concerned with the availability of nuclear energy, I should say by way of conclusion that the problem of energy availability may also be soluble by a technology that is not known today, but which is not entirely visionary, namely the fusion of hydrogen to helium.

There are three isotopes of hydrogen: the common form H (1p), deuterium D (1p, 1n), and tritium T (1p, 2n). The least difficult way to produce helium is through a combination of D and T. About one part in 6000 of the hydrogen in ordinary water is D (heavy water), so that a very large supply of D is available. But no natural supply of T exists (in fact T is radioactive, and disappears in a few tens of years to become the light isotope of helium). Hence T would need to be prepared artificially, either by adding a neutron to D, or by a reaction involving lithium. Neither of these processes seems likely to be economically useful, however. Lithium is a rare element, as can be seen from Figure 2.1, page 9, and so could have only a short lifetime if it were called on to supply T in quantity. Neutrons from uranium or thorium could convert D to T, but this would be to maintain nuclear fission as an essential process, and fusion would then not be worth the trouble (since fusion is much harder to achieve than fission). It is true that a neutron is released when D and T react together to produce the normal form of helium, He (2p, 2n), and that one might seek to use this neutron to make more T from D. But the efficiency of such a process would be low, at any rate for the reactor designs that have so far been discussed.

It seems then that fusion would need to follow the more difficult procedure of combining D with D. Whether this will ever prove to be economically feasible is still uncertain. A favourable outcome would be of great eventual importance, but meanwhile more immediate access to nuclear energy is urgently required, an access that can come at the present time *only* from fission.

References

W. B. Lewis, Elizabeth Laird Memorial Lecture, University of Western Ontario, 1974

W. B. Lewis, M. F. Duret, D. S. Craig, J. I. Veeder, and A. S. Bain, 'Large Scale Nuclear Energy from the Thorium Cycle', AECL-3980 Paper P/157, Proc. 4th U.N. Conf. P.U.A.E., Geneva, 1971

6
The Safety of Nuclear Energy

There is a good reason why nuclear energy is much the safest energy source. The safety comes from the fact that nuclear energy is not easily available. No individual can set up a privately owned nuclear reactor. Chemical energy, on the other hand, is freely available to us all, which is why chemical energy causes so many accidents. When I was a boy I even made gunpowder for myself. It was not difficult for a determined youngster to get hold of the necessary saltpeter, charcoal, and sulphur. Another boy, less cautious or less lucky than I was, might easily have lost a hand or an eye. Because familiarity breeds contempt, and because lack of familiarity breeds apprehension, we tend to think of unsafe chemical energy as being safe and to think of safe nuclear energy as being unsafe.

The New Zealand Alps look forbidding and very unsafe. Even the finest experts among mountaineers treat them with great respect. Yet I remember statistics of deaths from misadventure being published while I was visiting New Zealand. Deaths from mountaineering formed about 1½ per cent of the total; deaths in power boat accidents formed 23 per cent. I was surprised to find the one so low and the other so high, and this example taught me not to trust guesswork in assessing dangers of any kind. I doubt that anybody guesses instinctively the true danger of driving a car on the highway.

There are some people who simply do not want to know about nuclear energy. The association in their minds with nuclear bombs is strong, and they tend to think the two are really the same thing. Logically this attitude is not very sensible, any more than it would be sensible to say that because eating a piece of chocolate and exploding a hand grenade are both manifestations of chemical energy the two are the same.* I have talked with many worried people, and always at the back of their minds they have the concept that nuclear energy will lead to more bombs being made. It may, but only marginally so. Nothing concerning nuclear energy will prevent there being large stores of nuclear bombs in the world, for large stores exist already. It seems to me much more likely that these already-existing bombs will actually be used if the world runs desperately short of energy than if the world became energy-rich. So I would say that the development of nuclear energy is really a safeguard, probably an essential safeguard, against the outbreak of nuclear war.

The explosion some years ago of the *chemical* plant at Flixborough in Humberside gave those people in the U.K. who oppose the development of nuclear energy the chance of implanting in the public mind the image of exploding reactors flinging radioactive materials over the surrounding countryside. Because no *nuclear* reactor in the western democracies has in fact exploded, dark stories have been circulated of reactors coming desperately close to explosion, reactors rescued in the nick of time, like the perils of Pauline. But unlike the extensive death lists made public every few weeks after some unfortunate plane crash, no death lists or casualty lists remotely consistent with those supposed near-disasters have ever appeared. Nor will it do for those who spread these stories to imply a cover-up, for no civil authority

*** Both TNT and chocolate are made up from atoms of hydrogen, carbon, nitrogen, and oxygen. If one wanted to go to the trouble, the chocolate could be made into TNT.**

within the western democracies could possibly keep many deaths secret, even for days let alone years.

The strangest tale of this kind is of a Soviet reactor which exploded 18 years ago at a remote place in the Ural mountains. Lacking any real disaster in the West, a mysterious unconfirmed disaster is conjured out of an eastern hat. I might write a story myself about a man who decided to find out about that exploding reactor, who evaded the Soviet security machine, and then got himself to that remote place in the Ural mountains. But to do this actually myself would be out of my class. So I have no way to check this claim. First, I have no means of determining if an explosion really occurred. And if an explosion did occur, I have no means of determining whether the explosion was nuclear or chemical. (The story, as I heard it, told of a totally devastated area some 30 kilometres across, a bare desert of a place without vegetation. This sounds more like chemical poisoning than radioactivity.) But if there really was an explosion, and if the explosion really was nuclear, I would still like to know whether the explosion occurred in a civil rather than a military program. Military programs necessarily lack the safeguards of civil programs, since military programs are concerned with explosions *per se*. Because I have no way of deciding these matters, this tale remains for me a tale, as I think it should for every person who bases conclusions on verifiable facts.

At the very moment of writing, a leak of radioactive material has been reported from the nuclear installations at Windscale, quite close to my home in Cumbria. The media have given extensive coverage to this incident, but without mentioning the essential perspective that *everything* is radioactive. The food you eat, the air you breathe, and – God save us – both you and I are radioactive. The relevant question therefore for this leak is, *how much was it radioactive*? The media did not tell us. In a *Sunday Times* article (12

December 1976) a Windscale official was quoted as having said the leak was extremely minor, but since the article in question was headlined *How the nuclear cover-up went wrong* the reader was clearly intended to regard this official reassurance as part of a cover-up.

To speak precisely of *how much* radiation a person has suffered, for example through exposure to medical X-rays, a unit known as the *rad* has been defined. A chest X-ray involves us in an exposure which may be as low as 0.04 rad or as high as 1 rad. A gastronomic X-ray involves an exposure of about 1 rad, and X-ray exposures of the extremities range from 0.25 to 1 rad. A complete medical folio of X-rays involves a total exposure of about 25 rad.

Where forms of radiation other than X-rays and γ-rays are concerned, the unit of exposure is called the *rem*. An exposure of 1 rem produces closely the same damage to body cells as 1 rad of X-rays would do. (For our purpose, 1 rem = 1 rad.) The *maximum* permitted exposure for workers in the nuclear power industry is 5 rem per year, about equivalent to a significant X-ray examination although not nearly as much as a full folio of X-rays. (Doctors often take several exposures of parts of the body other than the part suspected of giving trouble, 'just to make sure'. A broken wrist which I had two years ago was thought to require 4 X-ray pictures, which must have cost me a total exposure of between 1 and 4 rad.) According to U.S. law, the maximum exposure which members of the public living close to a nuclear power plant may experience is $\frac{1}{200}$ rem per year, only about one-twentieth of what we experience all the time from the radioactivity of the rocks, of the soil, and even from the materials of which our houses are constructed. This natural radioactive background is usually about $\frac{1}{10}$ rem per year, but exceptionally in some parts of the world it is as high as 1 rem per year.

Because of this never-ending exposure to natural radioactivity, living creatures have developed mechanisms for repairing radiation damage to body cells. Unfortunately the repair facilities of the human body become inadequate for sudden exposures above 100 rem. Radiation sickness occurs, and survival becomes improbable above 400 rem. For sudden exposures below 10 rem, however, our repair facilities are good. This is why the maximum exposure for workers in the nuclear industry has been set at 5 rem per year, and why the medical profession gives X-ray examinations routinely up to about 5 rad.

With this background, let us return to the Windscale leak. A leak is significant only in so far as someone suffers exposure from it. The question therefore is what exposures did people at Windscale experience? It is to be hoped that this question will receive an answer and that, if the answer turns out to be only small, say $\frac{1}{100}$ rem, it is still given as much emphatic publicity as the story of the leak itself. The words 'extremely minor' suggest to me an exposure of this order, rather than the 1 rad that an abdominal X-ray would involve, and which, if it had happened to a member of the Windscale staff, would have received no publicity at all.

If the media had really been interested in the welfare of workers in the nuclear power industry, the sensible thing would have been to obtain and to publish mortality statistics which the United Kingdom Atomic Energy Authority (UKAEA) gathered over the period 1962–74. Table 6.1 gives these statistics.*

The remarkable fact is that in all categories actual deaths were significantly less than expected deaths. If the situation had been the other way round, with actual deaths significantly greater than expected deaths, every anti-nuclear environmentalist would have let loose a heart-rending cry, so loud indeed that the public would probably have become convinced of the 'case' against nuclear energy and attempted

* I obtained these numbers with no more trouble than a telephone call to the UKAEA information centre in London.

Table 6.1 *Mortality statistics; male employees of UKAEA and BNFL, 1962/1974*

Cause of death	*Actual number*	*Actuarially expected**	*Ratio: actual/expected*
All causes	2730	3652	0.75
All cancers	730	858	0.85
Circulatory system	1497	1651	0.91
Respiratory system	173	437	0.40
Digestive system	65	90	0.72
Genito-urinary system	26	47	0.55
Accidents, violence	153	197	0.78

* It would be actuarially expected for the same number of males from the same age distribution from the general population.

refutations of the statistics would have been brushed aside as a cover-up.

So how shall we interpret Table 6.1? Are we to believe that radiation is good for us, that it fights off serious diseases? One might reasonably argue that what is sauce for the goose is sauce for the gander, but supporters of nuclear energy have not sought to do this. Unlike their opponents, supporters of nuclear energy have never sought to misinterpret or to manipulate statistics. If exposure to some excess radiation were the only factor by which U.K.A.E.A. and B.N.F.L. workers differ from the general population, it would be right to regard Table 6.1 as a demonstration of the beneficial effects of moderate exposure to radiation. But the A.E.A. sites its power stations well away from heavily populated metropolitan areas. Windscale itself is close by the Lake District. So workers there may well take more healthy exercise than the general population does. Or the A.E.A. may give its staff better attention than the general medical services do; there might be other reasons as well.

It is ironic that anti-nuclear environmentalists show so much feigned apprehension for workers in the nuclear industry and so little for workers in other industries. Comparison of Table 6.1 with Table 6.2, especially for cancer

Table 6.2 *Mortality statistics*

Category	*Actual number*	*Actuarially expected*	*Ratio: actual/expected*	*Cause of death*
Coal-face workers	420	12	35.00	Respiratory
Stevedores, dock labourers	525	243	2.16	Respiratory
Deck and engine room ratings, barge and boatmen	344	162	2.12	Cancer
Constructional engineers, railway lengthmen, operators of earth-moving equipment, fishermen, window and office cleaners	647	133	4.86	Accidents (other than road and home)
Textile workers	238	137	1.74	Heart diseases
Publicans, innkeepers	617	281	2.20	Diabetes, nervous diseases
Company directors	425	54	7.87	Cancer, suicide, road accidents

From Registrar General's Supplement (1971)

and respiratory diseases, exposes the hollowness of the protestations of these self-styled concerned persons.

In the past few months I have given a few lectures in which I mentioned the need for nuclear energy. During the subsequent question period I have always been asked about the 'problem' of the disposal of radioactive wastes. There is no problem, and the thought that there is, if the concern is genuine, can only come from ignorance and from a fear of the unknown. A typical nuclear power plant would produce about 2 cubic yards of waste per year. Yes indeed, a mere 2 cubic yards. Those who argue against nuclear energy are always happy to imply waste on the scale of that from a coal-fired power station. Figure 6.1 shows the real situation. The total amount of nuclear waste produced in Britain* would be about 400 tons per year, which may be compared with the 120 000 000 tons of coal that we mine each year. Securing

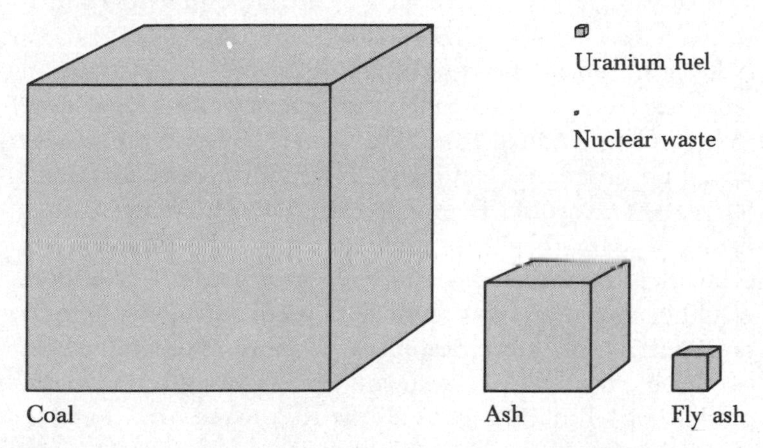

Figure 6.1 Relative fuel and waste volumes (after W. B. Lewis, *Energy in the Future of Man: From Survival to Super-Living*, University of Calgary, 1975)

storage space underground, say at a depth of 3000 feet, would evidently be a minuscule problem.

Anti-nuclear environmentalists seek to paint a disturbing picture of deeply buried nuclear waste surfacing again in the flow of water from springs and rivers. But there is a clear example to show that no such resurfacing actually takes place. The hills of the English Lake District generate within themselves as much radioactive energy as would come from the

* Assuming an all-nuclear economy.

buried waste products of three or four large nuclear power plants. This natural radioactivity, lying above sea level, inevitably has the heavy rains of the Lake District perculating downwards through it, not water that is imagined to rise miraculously upward from great depth. Yet experience shows that the natural radioactivity of the Lakeland hills does *not* wash out in our streams and rivers. This is a decisive, practical refutation of the claim of the anti-nuclear lobby.

The intensity of the radioactivity of waste falls off from year to year in a known way, which can be precisely calculated. A balance would be reached after about 50 years, a balance in which the rate of fall-off of the radioactivity of old waste became equal to the radioactivity of the new waste that was being added from year to year. Thus radioactivity would not increase indefinitely. It would increase for about 50 years, and would then stay essentially the same. In this balanced state the heat produced by all the radioactivity of all the accumulated waste lying at a depth of 3000 feet would be only a fraction of the heat produced by the natural radioactivity of the uranium and thorium contained in overlying rock. This statement applies to all the waste produced by Britain, and to all the rock taken over the area of Britain.

Most of the natural radioactivity which occurs all the time has no harmful effect. It is only the radioactivity occurring in a thin surface skin of soil which produces the radiation background that we ourselves experience. And waste buried at depth would likewise have no harmful effect. To guard against the possibility of its getting into solution, the radioactive waste will very likely be turned into a glassy material which would leech only very slowly should water ever reach it. And water in which radioactive materials might dissolve would typically require about 1000 years before any of it diffused upward to the surface. During this time harmful radioactivities would essentially disappear. It is certain, moreover, that radioactive atoms would leave the water on its way to the surface by entering and being trapped in the rock through which the water percolated. It has been calculated that, in the worst conceivable circumstance, the long-term effect of the accumulation of nuclear waste should not contribute to surface radioactivity more than 1 per cent of the effect of ordinary soil.

The imagined danger of radioactive waste was emphasized in the editorial which appeared 12 December 1976 in the *Sunday Times*. Since this editorial also illustrated quite a number of points which I have made in previous chapters, I will quote it at length:

> 'Must we go Nuclear?
> Is it lunacy to claim that we can do without nuclear energy altogether? Yes, it is, says the Atomic Energy Authority, with the passive support of Mr. Benn's Energy Department. The conventional wisdom in the power-generation game is that the United Kingdom's wealth in oil, natural gas and even coal will begin to dwindle after 1990, and that there is nothing worth talking about to fill the gap thus left except power from nuclear fission. We already derive about seven per cent of our electricity needs, comparatively cheaply, from that source. At present it comes from thermal reactors (using uranium). The Atomic Energy Authority – a hard-liner in this field – calculates that by the year 2000 the proportion will have risen to over 50 per cent, with nearly half of that supplied by fast breeder reactors (using plutonium bred from uranium).
>
> 'But that forecast supposes that the United Kingdom's energy demands, like its population, will go on growing as they have in the past. These expectations have proved as often wrong as right. We may lose our faith in the God-givenness of eternal cheap energy, and our consequent

lavish ways with it (just as we have at last begun to perceive that tap-water is not forever). Public bodies may devise techniques for getting unchanged results from reduced resources, as by the heating of whole districts with hot water from a single plant. Power engineers may improve their skill at drawing energy from the 'renewable' sources: sun, wind, tide and (in particular) waves.

'More than all that, nuclear power generation entails hazards to health and safety of a new magnitude. These are far more than the danger of radioactive leaks of the kind now disclosed at Windscale (with a tardiness at which Mr. Benn is rightly incensed). That threat could conceivably be lived with: other products too – coal, certain chemicals – are produced with great attendant risks. It is at least a danger which we would knowingly choose and face for ourselves. But nuclear power generation produces wastes which remain menacingly radioactive for hundreds of thousands of years. If fast breeder reactors became common, plutonium – an enormously toxic material – would be widely made, traded and transported. If could become an object of theft and an instrument of terrorism (as an atmospheric contaminant and even as the core of a home-made nuclear bomb). Alternatively, the means to prevent that might trespass grossly on personal liberties. In addition, the more countries know how to make nuclear electricity, the more are on the way to making nuclear bombs, and the sharper are the risks of nuclear warfare.

'These are therefore decisions of a special kind. Being irreversible, they affect the state of the world as far as the human mind can see and well beyond. A generation which took them without exhaustive thought about the consequences might stand convicted of a peculiarly wicked selfishness. Some of the problems can be lessened by careful organisation. Waste disposal can be coordinated by a government body. National and international safeguards can be devised against plutonium theft and the random spread of nuclear weaponry. But experience does not suggest that these arrangements can be made watertight; and public unease has not been allayed by the Government's disposition to "skulk behind the Official Secrets Act". (The phrase comes from Sir Brian Flowers, under whom the Royal Commission on Environmental Pollution has recently produced a powerful report urging nuclear caution.)

'We have time. We can at present make more electricity than we need. It is neither necessary nor economically possible to start work on the first commercial fast reactor within less than three years. During that time the behaviour of its prototype at Dounreay can be further studied. Techniques of waste disposal through vitrifaction can be advanced and tested. Research into renewable sources of energy can be pushed forward. National and international safety can be explored. And all this can be done in public. Mr. Shore, the Environment Secretary, has a good opportunity to make a start. He can call in the proposal for a £600m expansion of the Windscale plant – the plan which Cumbria County Council approved last month in ignorance of the leak – and order a public inquiry.'

When one looks through this seemingly persuasive editorial generally, not permitting oneself to be deflected by individual phrases or sentences, an overwhelming message stands out sharp and clear. The author's plain intention is to persuade us to stop or at least to delay the British nuclear program.

The last paragraph begins with an error. 'We have time.' We do not have time. It will be difficult to lay down all the capital investment necessary for our whole industry to shift from hydrocarbon energy to nuclear energy even if we go immediately ahead with the greatest determination. This 'We have time' must be a deliberate mistake, because every knowledgeable person in the field is well-aware that time is desperately short, particularly in a world-wide perspective.

I remarked already in Chapter 5 on the anti-nuclear tactic of using the admitted problems which still exist with fast plutonium breeders to discredit all nuclear energy sources. I spoke about how the anti-nuclear campaign avoids mention of entirely viable alternatives, like the CANDU-system. Here you can see this tactic in full operation.

Here too we have the attempt to frighten people with the scare of long-lived radioactivities:

> 'But nuclear power generation produces wastes which remain menacingly radioactive for hundreds of thousands of years.'

It doesn't. The wastes which would build up in hundreds of thousands of years of operation would be no more menacingly radioactive than the wastes which build up in ten years, and the wastes which build up in ten years are no more menacingly radioactive than the radium, and the decay products of radium, contained in the fly ash produced by coal-fired power plants.

This matter is worth driving home at some length. Let us consider more closely what radioactive waste really is. The pieces into which uranium and thorium break up under fission are nearly all radioactive. That is to say, they undergo nuclear processes themselves which in the long run change them into atoms that are stable. A large variety of pieces is produced, because uranium (or plutonium) does not by any means always go into the same two fragments when it undergoes fission. Fission is analogous to the break-up of water drops, and there are variations of detail from one nucleus to another, which lead to about a hundred different kinds of piece being produced. Almost all are radioactive, but different pieces have their own individual forms of bchaviour.

One important individual characteristic is the *half-life* of the radioactivity of each piece. The half-life has the following meaning. Start with a large number of pieces of the same kind. After a time interval equal to the half-life, one-half (very nearly) of them will have undergone their particular form of radioactivity, being changed either into a stable atom or into an atom with a different form of radioactivity. And after a second time interval equal to the half-life, one-half of those pieces that were unchanged after the first interval will now have changed, so that only one-quarter of the original pieces remain. And so on for additional intervals of time. Among the many different pieces are some with half-lives of less than a second, some with half-lives of several seconds, tens of seconds, minutes, hours, days, years, and tens of years. Then the more or less smooth range of half-lives stops and there is a big gap, with the next largest half-lives in the hundreds of thousands of years. The fission fragment Tc-99 has a half-life of about 200 000 years. The symbol Tc stands for the atom technetium, and Tc-99 is the isotope of technetium with 43 protons and 56 neutrons, Tc (43p, 56n). Some 2 to 3 per cent of the weight of uranium or plutonium undergoing fission becomes Tc-99.

Sr-90 (the isotope of strontium with 38 protons and 52 neutrons), and Cs-137 (the isotope of cesium with 55 protons and 82 neutrons), are the most troublesome components of radioactive waste. Sr-90 has a half-life of about 28 years, and Cs-137 a half-life of 30 years, and both also comprise 2 to 3 per cent of the uranium or plutonium undergoing fission. Lct us compare the problem of Tc-99 accumulation in nuclear wastes with the problem of the accumulation of Sr-90.

Suppose we were to build a stockpile of Sr-90 at a rate of 1 gram per year. For a while the stockpile would increase, but eventually the stockpile would become so large that the loss of Sr-90 through radioactivity (radioactive decay as it is often called) would equal the rate at which new Sr-90 was being added. Calculation shows that this balance occurs when the stockpile contains about 40 grams of Sr-90. Thereafter, no matter how long we go on adding our 1 gram per year, the amount of Sr-90 in the stockpile never gets any bigger.

Now for each gram of Sr-90 in the waste we also have about 1 gram of Tc-99. Suppose we build a similar stockpile for Tc-99. Because of the much longer half-life of Tc-99 we shall be able to go on increasing its stockpile long after the Sr-90 in the stockpile has reached its steady balance of 40 grams. In fact, we shall be able to build a stockpile of Tc-99 10 000 times larger just because the Tc-99 half-life is 10 000 times longer. The balance for Tc-99 will therefore occur at 400 000 grams.

If we are to please the anti-nuclear lobby we must throw up our hands at this point, stop thinking, and cry 'too much'. But if we do, we shall have been taken in by a deception. Although the mass in the Tc-99 stockpile is much bigger than that in the Sr-90 stockpile, the radioactivity is not. To obtain a decay rate equal to that of 1 gram of Sr-90 it takes 10 000 grams of Tc-99, so the 400 000 gram stockpile of Tc-99 produces just the same number of radioactive decays as the 40 gram stockpile of Sr-90. Indeed, the decays of Tc-99 are less energetic than those of Sr-90, so the Tc-99 stockpile would actually produce a less harmful radiation background than the Sr-90 stockpile.

While one might conceivably argue that this deception came from technical ignorance, there is a second deception which cannot be so innocently attributed. If the writer of the *Sunday Times* editorial were to be proved correct, if an adequate supply of energy were to become available from Sun, wind, and (miraculously) waves, the Tc-99 stockpile would never remotely approach its balanced condition.

Society will only accumulate Tc-99 for hundreds of thousands of years if nuclear fission proves to be necessary for hundreds of thousands of years. Should a superior technology for providing ample solar energy be discovered a few decades hence, society would stop accumulating Tc-99. Would the waste accumulated over those decades really be menacing? Stored underground, the Tc-99 could hardly add more than one part in 100 000 to the truly long-lived natural background of radiation from ordinary rock and soil.* Is this an adequate reason for blocking access to the energy which the world will need so desperately a decade or two hence? The real menace comes of course not from Tc-99, but from the editorial in the *Sunday Times.*

Although nuclear reactors have proved themselves to be exceedingly safe, very much safer than other sources of energy, in a world committed to nuclear energy, accidents would happen occasionally. What would a serious accident involve? There would be an escape of radioactive materials into the environment. People close to the accident would suffer exposure to radiation, but the life expectancy of each individual would not be much impaired. The situation would be about like the exposure you would undergo if you had several X-rays taken, say for an arthritic hip joint. An exposure worse than this would be almost impossibly rare – a worse situation might happen once in a million years.

Although the life expectancy of each individual would not be much impaired, if you add up many small probabilities of unnatural death you arrive at a number of impersonal, expected, 'statistical deaths'. Calculations of this kind have been made in what is called a 'conservative' manner, which is to say the number of these statistical deaths has been overestimated, perhaps grossly so. To see where this point lies, consider the following example.

* The natural background will last, not just for hundreds of thousands of years, but for thousands of millions of years.

For two groups of people, the second group is a thousand times more numerous than the first group. An accident causes each member of the first group to suffer an exposure of 5 rem, and a second accident causes each member of the second group to experience an exposure of $\frac{1}{200}$ rem. This makes the total radiation exposure for each group the same. Will the number of 'statistical deaths' also be the same? If cancer arises from damage to an individual body cell the answer to this question would be, yes. But if cancer develops from an association of several damaged cells, if it is multi-celled in origin, the answer is, no. In the latter case the group with the lower exposure will suffer many fewer statistical deaths.

The evidence favours the multi-cell possibility, but the evidence has not yet been judged decisive by health-physics experts. So what is done in making 'conservative' calculations is to assume the single-cell possibility, even though the weight of the evidence is against it. Not all authoritative bodies agree with this conservative procedure, because it opens the door to scaremongers whose prime concern is to frighten people. The United Nations Scientific Committee on Effects of Atomic Radiation is an example of a body which has not

Table 6.3 *Conservatively estimated annual cancer deaths due to radiation from aspects of a U.S. nuclear energy industry generating 3.5×10^{12}kWh of electricity**

Source	*Deaths/year*
Routine emissions	8.7
Reactor accidents	10.0 (600)†
Transportation accidents	0.01
Waste disposal	0.4
Plutonium (routine release)	0.1
Total	20.00 (600)†

* Does not include effects of long half-life radioactivities, sabotage, or terrorism.

† Numbers in parentheses represent worst claim by critics of the nuclear energy industry.

From Bernard L. Cohen, 'Impacts of the Nuclear Energy Industry on Human Health and Society', American Scientist, *September–October 1976, 552* endorsed this conservative way of making statistical calculations.

Table 6.3 gives the conservatively calculated statistical deaths from the listed causes for the United States to be expected if the U.S. were to derive all its presently used energy from nuclear power stations. The table is taken from a recent article by Bernard L. Cohen (*American Scientist*, September–October 1976). Professor Cohen compiled the table from data given in a multi-volumed report to the U.S. Atomic Energy Commission, a report known as WASH-1400. Professor Cohen's table also shows an environmentalist claim that the number of statistical deaths from accidents to reactors should be considerably more than the number calculated in WASH-1400. This claim comes from a difference in assessing small chances for what are called 'meltdown' accidents.

Professor Cohen uses the estimates of Table 6.3 to make a number of ironic comparisons; I shall end here by quoting them:

> Some Statistically Comparable Risks
>
> Table 1 (6.3 here) reveals that estimates based on acceptance of WASH-1400 predict that an all-nuclear energy economy would result in about 20 deaths per year; critics of nuclear energy claim that the number is about 600. I shall now attempt to put these estimates in perspective. Since cancer is delayed by 15 to 45 years after exposure, the average loss of life expectancy per victim is 20 years. The loss of life expectancy for the average American from these 20 deaths per year is then (20×20 man-years lost/2×10^8 man-years lived) $= 2 \times 10^{-6}$ of a life-time $=$ 1.2 hours.
>
> Some of us subject ourselves to many other risks that reduce our life expectancy, such as smoking cigarettes. One pack per day (3.6×10^5 cigarettes) reduces life

expectancy by about 8 years (36) which, assuming linearity, corresponds to 12 minutes loss of life expectancy per cigarette smoked; thus the risk of nuclear power is equivalent to that of smoking 6 cigarettes in one's lifetime, or one every 10 years.

Statistics show that life expectancy in large cities is 5 years less than in rural areas (37). This phenomenon may be explained in part by differences in racial makeup, but it is believed to be largely due to the strains of city life. If linearity is assumed, the risk of nuclear power (1.2 hr loss of life expectancy) is equal to the risk of spending 16 hours of one's life in a city.

Riding in automobiles exposes us to a death risk of 2×10^{-8} per mile, or if 35 years of life are lost in an average traffic fatality, loss of life expectancy is 7×10^{-7} years per mile travelled. The risk of nuclear power is then equal to that of riding in automobiles an extra 3 miles per year.

Riding in a small rather than a large car doubles one's risk (38) of fatal injury, so the 1.2 hours loss of life expectancy from all nuclear power is equivalent to the risk of riding the same amount as at present, but 3 miles per year of it in a small rather than a large car.

Another risk some of us take is being overweight. If we assume loss of life expectancy to be linear with overweight, the 1.2 hour loss from all-nuclear power is equivalent to the risk of being 0.02 ounces overweight (39). However, Pauling (40) has shown that the data are better fit by a quadratic dependence, and if this is accepted the risk of all-nuclear power is equivalent to that of being 0.3 pounds overweight.

All these estimates are based on the government agency projection of 20 deaths per year. If we instead accept the critics' estimate of 30 times as many deaths, the risk of nuclear power is equivalent to the risks of smoking 3 cigarettes per year, spending 20 days of one's life in a city (one day every three years), riding in automobiles an extra hundred miles per year, riding in automobiles the same amount as at present but 100 miles of it per year (1%) in a small rather than a large car, or being 0.6 ounces overweight on the linear hypothesis or 1.6 pounds overweight on the quadratic hypothesis.

Additional perspective may be gained by comparing the effects of an industry deriving electric power from coal by present technology. The most important environmental impact of coal-fired power is air pollution, which, it is estimated, would cause about 10,000 deaths per year (41), at least an order of magnitude more than even the critics estimate would be caused by nuclear power. In addition, this air pollution would cause (41) about 25 million cases per year of chronic respiratory disease, 200 million person-days of aggravated heart-lung disease symptoms, and about $5 billion worth of property damage; there are no comparable nuclear risks. And the mining of coal to produce this power would cause about 750 deaths (42) per year among coal miners, more than 10 times the toll from uranium-mining for nuclear power, and the uranium figure would be reduced about 50-fold with breeder reactors. Transporting coal would cause about 500 deaths per year (43), two orders of magnitude more than would be caused by transportation for the nuclear industry.

It is important to point out that the numbers in Table I (6.3 here) (and the pespective I have put them in) are based on annual averages. As the critics are constantly reminding us, if their estimates are correct, there might be an accident every ten years with several thousand deaths and every 50 years with tens of thousands of deaths. It is not difficult to make this prospect seem extremely dismal. On the other hand we should not envision these accidents as producing stacks of dead bodies; the great majority of fatalities predicted are from cancer that would occur 15 to 45 years later. In nearly all cases, the affected individuals would have only about a 0.5% increased chance of getting cancer. The average American's risk of cancer death is now 16.8%; typically it would be increased to 17.3%. For comparison, the average risk varies from 18.4% in New England to 14.7% in the Southeast, but these variations are rarely noticed.

If an area were affected by a nuclear accident and authorities revealed that, as a result, the average citizen's probability of eventually dying of cancer was increased from 16.8 to 17.3%, it would hardly start a panic. We have had some experience with a similar but much more serious situation: when reports first reached the public of the risk in cigarette-smoking, tens of millions of Americans were suddenly informed that they had accrued a 10% increased probability of cancer death, 20 times larger than the effects from a nuclear accident. Even that story did not stay in the headlines long, and it brought very little counteraction.

Critics often raise the point that the risks of nuclear power are not shared equally by all who benefit but are disproportionately great for people who live close to a nuclear power plant. This, of course, is true for all technology, but let us put it in perspective. The risk is 10^{-6} per year if we accept the WASH-1400 accident estimates, or equal to the risk of riding in an automobile an extra 50 miles per year or 250 yards per day.

Thus if moving away from a nuclear power plant increases one's commuting distance by more than 125 yards (half a block), it is safer to live next to the power plant. If one prefers the estimates we attribute to the critics (20 times larger), it does not pay to move away if doing so increases commuting distances by more than 1.5 miles per day. Even with the critics' estimates, living next to a nuclear power plant reduces life expectancy by only 0.03 years, which makes it 150 times safer than living in a city.

References

(34) J. McPhee, *The Curve of Binding Energy* (New York: Farrar, Strauss, and Giroux, 1974)

(35) T. B. Taylor, at Hearings of Joint Committee on Atomic Energy, June 1975

(36) U.S. Surgeon-General, 1962. Report on cigarette smoking

(37) E. Teller and A. L. Latter, *Our Nuclear Future*, page 124 (New York: Criterion Books, 1958)

(38) Insurance Institute for Highway Safety, vol. 10, no. 12 (9 July 1975), gives 24.6 fatalities per 100 000 vehicle years registered for small cars and 11.3 for large cars. A similar conclusion is obtained from National Safety Council, *Accident Facts–1965* (Chicago)

(39) Being overweight by 25% (about 40 lb for the average adult male) reduces life expectancy by about 5 years (L. I. Dublin and H. H. Marks, *Mortality among Insured Overweights in Recent Years*, New York: Metropolitan Life Insurance Co., 1952.) Since 1.2 hours is 1.4×10^{-4} years, a linear hypothesis gives the risk to be that of 8 lb/yr $\times 1.4 \times 10^{-4}$ yr $= 1.2 \times 10^{-3}$ lb $= 0.02$ oz.

(40) L. Pauling, *Proc. Nat. Acad. Sci.*, 1958, **44**, 619

(41) U.S. Senate Committee on Public Works, 1975, 'Air quality and stationary source emission control', gives values for an urban plant (p. 631) and a remote plant (p. 626); we use the average of these and multiply by 400 for the number of plants. A larger number of deaths is estimated by R. Wilson, paper presented at Energy Symposium, Boulder, CO, June 1974

(42) B. L. Cohen, *Nuclear Science and Society*, pp. 139–40 (New York: Doubleday-Anchor, 1974)

(43) L. Sagan, *Nature*, 1974, **250**, 109

American Scientist, Volume 64, 1976 September–October

Index